WATER RESOURCE PLANNING, DEVELOPMENT AND MANAGEMENT

WATERSHEDS

PROCESSES, MANAGEMENT AND IMPACT

WATER RESOURCE PLANNING, DEVELOPMENT AND MANAGEMENT

Additional books in this series can be found on Nova's website under the Series tab.

Additional E-books in this series can be found on Nova's website under the E-book tab.

ENVIRONMENTAL SCIENCE, ENGINEERING AND TECHNOLOGY

Additional books in this series can be found on Nova's website under the Series tab.

Additional E-books in this series can be found on Nova's website under the E-book tab.

WATER RESOURCE PLANNING, DEVELOPMENT AND MANAGEMENT

WATERSHEDS

PROCESSES, MANAGEMENT AND IMPACT

GAVYN L. KUHN
AND
JASON R. EMERY
EDITORS

Nova Science Publishers, Inc.
New York

For permission to use material from this book please contact us:
Telephone 631-231-7269; Fax 631-231-8175
Web Site: http://www.novapublishers.com

NOTICE TO THE READER

The Publisher has taken reasonable care in the preparation of this book, but makes no expressed or implied warranty of any kind and assumes no responsibility for any errors or omissions. No liability is assumed for incidental or consequential damages in connection with or arising out of information contained in this book. The Publisher shall not be liable for any special, consequential, or exemplary damages resulting, in whole or in part, from the readers' use of, or reliance upon, this material. Any parts of this book based on government reports are so indicated and copyright is claimed for those parts to the extent applicable to compilations of such works.

Independent verification should be sought for any data, advice or recommendations contained in this book. In addition, no responsibility is assumed by the publisher for any injury and/or damage to persons or property arising from any methods, products, instructions, ideas or otherwise contained in this publication.

This publication is designed to provide accurate and authoritative information with regard to the subject matter covered herein. It is sold with the clear understanding that the Publisher is not engaged in rendering legal or any other professional services. If legal or any other expert assistance is required, the services of a competent person should be sought. FROM A DECLARATION OF PARTICIPANTS JOINTLY ADOPTED BY A COMMITTEE OF THE AMERICAN BAR ASSOCIATION AND A COMMITTEE OF PUBLISHERS.

Additional color graphics may be available in the e-book version of this book.

LIBRARY OF CONGRESS CATALOGING-IN-PUBLICATION DATA

Watersheds : processes, management, and impact / editors, Gavyn L. Kuhn and Jason R. Emery.
p. cm.
Includes index.
ISBN 978-1-62081-057-6 (hardcover)
1. Watersheds. 2. Watershed management. 3. Watersheds--Environmental aspects. 4. Watersheds--Research. I. Kuhn, Gavyn L. II. Emery, Jason R.
GB980.W38 2011
551.48--dc23

2012004475

Published by Nova Science Publishers, Inc. † New York

CONTENTS

PREFACE

This book examines the management and impacts of watersheds on the ecosystem process. The authors present topics including the effects of thinning on material transformation in a stand of Cryptomeria japonica in a watershed in eastern Japan; natural and human influences on the inorganic water chemistry of the northern Xinjiang rivers in China; current land use within the Conecuh/Escambia River watershed in the Gulf of Mexico; factors affecting farmers' adoption of best management practices in Ontario's Grand River watershed; and the adaptation of national elevation dataset to estimate depression storage in a glaciated watershed.

Chapter 1 - In a watershed in eastern Japan, seasonal changes of nutrient transformation were investigated in a stand of *Cryptomeria japonica*, an important coniferous tree in the cultivated forest of Japan. At first the authors described the high activity of nitrogen fixation, exceptionally in the litter of *C. japonica*. Patterns of denitrification and nitrification differed between lower and upper parts of the slope in the watershed. Seasonal changes of input and output of trace metals (cadmium, lead) and their distribution were also described. The thinning practice apparently saw the accumulation rates of aboveground C and N unaltered, although the contents and deposition rates of aboveground C and N were decreased by thinning. Nitrogen in green needles returned by thinning have a function whereby the N is temporarily stored and released later because the N immobilization and N fixation dominated compared to decomposition of the needles at this stage.

Chapter 2 - The influence of rock mineralogy, climate, relief and human activity on natural water composition and export as a result of weathering is a major scientific concern both at the local scale, because of the link between water chemistry and biodiversity, and at the global scale, in relation to global climate changes. This study evaluates the inorganic water qualities of the northern Xinjiang rivers in order to assess the natural and anthropogenic factors that have influenced the inorganic water chemistry of natural water in arid environment. Wide spatial variations are geographically characterized in the dissolved solids (TDS) and water chemistry. A typical carbonate alkaline with Ca^{2+}, HCO_3^- and SO_4^{2-} ions dominated the major inorganic water composition. The water chemistry agrees well with the "rock dominance" mechanism with a TDS value of 80-600 mg/L and a Na/(Na+Ca) ratio of 0.1-0.6. Stoichiometric analyses suggest that most of the ions derive from multiple sources. Carbonate weathering is the primary source, followed by silicate weathering and evaporite dissolution. The effects of local pollution have somewhat greater contribution on the oases and central areas of the Zhungarer watershed. Most parts of rivers show a trend toward an increase in the dissolved load from the upper to lower reaches, which is primarily attributed to evaporation effect of climate along the water course. Variations and distributional patterns of the inorganic water chemistry show evident correlation with regional lithological distribution, terrain and precipitation gradients, indicating a strong influence of local geology, topography and climate. Using rainwater as a baseline, contributions from atmospheric precipitation and rock weathering to the tributary chemistry are roughly evaluated to be in the range of 2-39% (average 13%) and 59-98% (average 86%), respectively.

Chapter 3 - The Gulf of Mexico sturgeon *Acipenser oxyrinchus desotoi* is an anadromous subspecies of the Atlantic sturgeon *Acipenser oxyrinchus oxyrinchus*. This federally threatened fish once inhabited Gulf coastal rivers from the Mississippi River to Tampa Bay, but is now rarely found south of the Suwannee River in Florida. The Conecuh/Escambia River system of Alabama and Florida contains more potential Gulf sturgeon spawning habitat than any other Florida Panhandle river, yet limited information is known about the sturgeons of this river. Radio transmitters were placed in 14 Gulf sturgeons collected from Escambia Bay during March through May of 2005. These radio-telemetered Gulf sturgeons were manually tracked via boat and radio receivers during April through November 2005 to collect summer holding area and movement data. Two, possibly three, summer holding areas were identified during this study, two in the Escambia River and one in the Delaney River.

Chapter 4 - While there are a number of possible causes including greenhouse gases, solar radiation, land use changes that may contribute to the changes in the worldwide climate, the reality is that regardless of the causes, much of the scientific community is in agreement that climate change is real. The global climate change during the recent years has been considered as one of the biggest threats to Florida. Many factors including solar radiation, atmospheric composition, cloud cover, ocean currents, and land activities influence the climate and make long-term changes. Florida's topography and its warm climate make the state vulnerable to the effects of climate change. Although Florida has abundant sources of surface water and groundwater, it is reported that water levels in Florida have dropped at an average rate of 0.10 to 0.15 m per year since 1950. This review article examines the impact of global climate change on Florida, especially on its watersheds, and potential climate change adaptation strategies.

Chapter 5 - To see how recent regulations and programs have operated in the Grand River watershed, the authors talked to Conservation Authority officials and consulted recent water quality studies to identify the sub-watersheds with the worst and best water quality. This chapter describes how residents within the sub-watersheds of the Grand, the Eramosa/Speed (E/S) and the Canagagigue Creek (CC) perceive agri-environmental programming and recent legislation imposed after the Walkerton tragedy. Earlier research had found fewer water quality problems in the E/S region than in CC. The authors thought this might be due to different land use practices, degrees of agricultural intensity, the attitudes of rural people toward environmental best management practices (BMPs) and the impact those practices might have on rural people's perceived quality of life.

Chapter 6 - The construction of raised roads may modify the hydraulic connectivity of natural depression storages (e.g., wetlands) and create additional artificial storages. As a result, it would noticeably alter the natural hydrology of the inclusive watershed. Quantifying these storages is very important for watershed management that aims to increase water supply and mitigate flooding. This can be done by processing a digital elevation model (DEM) in a geographic information system (e.g., ArcGIS®) on the basis of the U.S. Geological Survey's 1.61-km by 1.61-km quadrangle sections. However, DEMs with a spatial resolution finer than 30 m are not available for most watersheds. The objective of this study was to examine whether and how the 30-m National Elevation Dataset can be adapted to estimate potential depression storage capacity (PDSC), the total volume below the lowest point (i.e., breach) of the roads surrounding a section. The study was conducted in the Forest River watershed located in northeastern North Dakota. The results indicate that the estimation of PDSC tended to have larger errors for sections with a higher topographic relief than for sections with a lower relief. In addition, the estimation was sensitive to errors in the breach elevation values. Nevertheless, the PDSC for a section within the study watershed can be estimated as the multiplication of the value computed using the NED data by an adjustment coefficient of 1.05 to 1.3. One reasonable generalization of this study is that NED can be adapted to estimate PDSCs for watersheds where high-resolution DEMs are not available.

In: Watersheds
Editors: G. L. Kuhn and J. R. Emery
ISBN 978-1-62081-057-6

Chapter 1

EFFECTS OF THINNING ON MATERIAL TRANSFORMATION IN *CRYPTOMERIA JAPONICA* STAND IN A WATERSHED IN EASTERN JAPAN

Takashi Yamanaka*[*,1], *Keizo Hirai*[2], *Yuko Itoh*[1], *Tomoaki Morishita*[3], *Kyotaro Noguchi*[3] and *Shuichiro Yoshinaga*[4]

[1]Forestry and Forest Products Research Institute, Tsukuba, Ibaraki, Japan
[2]Tohoku Research Center, Forestry and Forest Products Research Institute, Morioka, Iwate, Japan
[3]Shikoku Research Center, Forestry and Forest Products Research Institute, Kochi, Kochi, Japan
[4]Kyushu Research Center, Forestry and Forest Products Research Institute, Kumamoto, Kumamoto, Japan

ABSTRACT

In a watershed in eastern Japan, seasonal changes of nutrient transformation were investigated in a stand of *Cryptomeria japonica*, an

[*] Corresponding autor: Takashi Yamanaka.

important coniferous tree in the cultivated forest of Japan. At first we described the high activity of nitrogen fixation, exceptionally in the litter of *C. japonica*. Patterns of denitrification and nitrification differed between lower and upper parts of the slope in the watershed. Seasonal changes of input and output of trace metals (cadmium, lead) and their distribution were also described. The thinning practice apparently saw the accumulation rates of aboveground C and N unaltered, although the contents and deposition rates of aboveground C and N were decreased by thinning. Nitrogen in green needles returned by thinning have a function whereby the N is temporarily stored and released later because the N immobilization and N fixation dominated compared to decomposition of the needles at this stage.

1. Introduction

The nutrient cycle occupies a key position in ecosystem processes. With some exceptions, nutrients are present far below tree-growth requirements, and productivity is closely tied to nutrient cycling. The majority of nutrients required for plant growth are derived from dead organic matter released by the microbes and animals composing the detrital food chain within local ecosystems. However, inputs from external sources are also important because they replenish nutrients that are lost from the system. Nutrients are input to the ecosystem via three primary channels: atmospheric deposition, weathering of primary minerals, and biological N_2 fixation.

An ecological system has various processes of energy and matter under variable environmental conditions such as soil type and climate. Considerable research has been done on each of the ecological processes; however, there has been little information concerning the relationships and functions of those that maintain the ecosystems (Likens and Bormann, 1995), making it difficult to evaluate the impact of human activity on forest resources.

In Japan, *Cryptomeria japonica* D. Don, an important coniferous tree that occupies 48% of the cultivated forest of Japan, has been planted for wood production in Japan, and is now ready to be harvested. Accordingly, we need information concerning the effect of thinning on the soil nutrient dynamics for reforestation after harvesting, biodiversity conservation, and sustainable forest management.

In order to obtain this information, which is usually lacking for entire ecological systems, members of Department of Forest Site Environment, Forestry and Forest Products Research Institute, Japan conducted a research project on the quantitative evaluation of the flux and cycling of water and N in a small watershed at Katsura, Shirosato, Ibaraki Prefecture, Japan. Subsequently, various trace elements were included in the measurements. In this chapter, we will present studies focusing on certain aspects of cycling and the input of nutrients which was carried out at this research site.

The research site was established at the Katsura Headwater Catchment (KHC) located about 120 km northeast of Tokyo, Japan (36°32'N, 140°18'E; 210-270 m a.s.l). The stream of this catchment joined with Naka River and discharges into the Pacific Ocean. The site, measuring 2.3 ha, is underlain by consolidated shale and sandstone of Mesozoic strata and overlain by a thick volcanic ash soil. The topography of the KHC consists of steep (>30°) valley side slopes, with a gentle incline on the upper parts. The climate is temperate. The rainfall is distributed fairly evenly year-round. At Kasama City, the nearest meteorological station is situated about 15 km from the study site, recording mean annual precipitation and temperature of 1330 mm and 13.1° (1971-2000), respectively. The lower and middle parts (1.0 ha) of the KHC are covered by a 45-year-old *C. japonica* stand (CS), whereas the upper part is covered by a deciduous broad-leaved forest (DBF), where *Quercus serrata* Murray, *Prunus Jamasakura* Sieb .ex. Koidzumi and *Carpinus laxiflora* (Sieb. et Zucc.) Blume dominate (Hirai et al., 2007).

In the following sections, we describe firstly seasonal changes in nutrient dynamics, and then effects on thinning on nutrient cycling.

2. Seasonal Changes of Material Transformation

2.1. Nitrogen Fixation

Biological N_2 fixation as well as atmospheric deposition is a major N input that affects the productivity of the forest ecosystem, but little attention has been paid to this process in woody debris due to its relatively low productivity compared with symbiotic N_2 fixation. In the Pacific Northwest of the United States, N_2 fixation rates in actinorhizal symbiosis were reported to be 23–320 kgN ha^{-1} yr^{-1} in an *Alnus rubra* Bong. forest (Hibbs and Cromack, 1990), whereas asymbiotic N_2 fixation in woody debris is estimated to be mostly below 1.0 kgN ha^{-1} yr^{-1} in a wide range of ecosystems (Sollins et al., 1987). Nevertheless, this input may play a significant role in the long-term N dynamics of the site, because in the course of plant succession, leguminous or actinorhizal plants, which are responsible for symbiotic N_2 fixation, become shaded and biological N_2 fixation is likely restricted to soil, litter, or logs, where N_2 is fixed asymbiotically (Perry, 1994).

In Japan, several studies on N_2 fixation during needle- or leaf-litter decomposition have been reported (Nioh, 1980; Shima and Chiba, 2002). *Cryptomeria japonica*, an important coniferous tree that occupies 48% of the cultivated forest of Japan, showed N_2-fixing activity at an early stage of needle-litter decomposition, which is referred to as F1 fraction (Nioh, 1980) and described as a much darker litter.

In this section we report the effects of the decomposition stage and site conditions on the N_2-fixing activity of *C. japonica* litter, and compare the activity with that of two other species, *Pinus densiflora* Sieb. et Zucc. and *Q. serrata*, which commonly grow in secondary forests. We measured the N_2-fixing activity, decomposition rate, N content, and fungal biomass of *C. japonica*, *P. densiflora* and *Q. serrata* litter incubated in litterbags set at different positions of the same slope of KHC.

Newly fallen litter of these three species was air-dried in the laboratory for one week, and three grams of air-dried litter were enclosed in a litterbag. After removal of the newly fallen litter, the litterbags were placed on the forest floor, then collected after 3, 16, 19, 31 and 34 months of incubation and used to measure N_2-fixing activity, N concentration and fungal biomass. The N_2-fixating activity of woody debris was measured using the acetylene reduction (AR) technique (Yamanaka et al., 2011). Fungal biomass was determined by the content of ergosterol, which is a component of fungal hypha membranes (Weete, 1974), and therefore should be correlated with the metabolically active biomass of a fungus (Nylund and Wallander, 1992). Total N content of the dried samples was measured using an NC analyzer (Sumigraph NC-900; Sumika Chemical Analysis Service, Osaka, Japan).

The AR activity in *Cryptomeria* litter exceptionally exceeded that of various other species. Nioh (1980) found that the AR activity peaked at 14.8 (nmoles $C_2H_4 \cdot h^{-1} \cdot g^{-1}$ DW) in the F1 layer of *Cryptomeria* litter. Yamashita et al. (1998) reported that AR activity in *Cryptomeria* litter was 25 after 12 months of incubation. In the present study, the AR activity in *Cryptomeria* litter in DBF increased to 62.65 after 19 months of incubation (Figure 1), while that in the litter of *Pinus* and *Quercus* increased to 1.07 and 0.62 respectively. Baker and Attiwill (1984) reported AR activity in the litter of *Pinus* and *Eucalyptus* at 2.47 and 1.36, respectively, while O'Connell et al. (1979) showed that AR activity in *Eucalyptus* leaf litter was 3.75.

The AR activity in *Cryptomeria* litter may be attributed to the chemical composition of decomposing litter (Nioh, 1989). In the decomposing litter of

C. japonica, the concentrations of water soluble potassium, calcium and magnesium increased, and the pH value of the litter increased from 5.0 in fresh litter to 7.9 in black litter (Takahashi, 1996). Conversely, the pH value in the F and H layers of *Pinus* and *Quercus* litter were more acidic than those in fresh litter (Ovington and Madgwick, 1957). Some azotobacters, the main members of aerobic free-living N_2-fixing bacteria, are known to grow in neutral to slightly alkaline conditions (Mulder and Brotonegoro, 1974). These results show that N_2-fixing bacteria grow well and fix N_2 restrictedly in decomposing black *Cryptomeria* litter. The bacteria contributing to N_2 fixation in the decomposition of *C. japonica* litter should be identified to clarify their distribution and function in decomposing *Cryptomeria* litter.

The ratio of N content to initial N content during decomposition was higher in *Cryptomeria* than in *Pinus* and *Quercus* (Table 1). Subsequently, changes in N content in *Cryptomeria* litter showed the same trend as that in *Quercus* and *Pinus* litters. In a litterbag study, Yamashita et al. (1998) reported that N mineralization and nitrification in litter of *C. japonica* accelerated after AR activity increased between 6 and 12 months of incubation. Nioh (1980) showed that nitrification was active in the F2 and H layers while N_2-fixing activity was high in the F1 fraction. These results suggest that, while N content in *Cryptomeria* litter characteristically increased, subsequent changes in the N content in *Cryptomeria* litter during decomposition advanced similarly to *Quercus* and *Pinus*, in terms of the balance of gain and loss by N mineralization, nitrification or denitrification.

During the initial 3 months, the ergosterol content in *Cryptomeria* and *Quercus* litter was higher than that in *Pinus* litter (Table 2). Ergosterol is the predominant fungal sterol, and its content is a sensitive indicator of fungal biomass (West et al., 1987). Therefore, this result indicates that the litter of *Cryptomeria* and *Quercus* was well-colonized by fungi at this stage. Then, ergosterol content in *Cryptomeria* remained the highest, whereas that in *Quercus* was the lowest among the three species after 19 months of incubation. The difference in ergosterol content in the litter of these two species matches the difference in their decomposition constant. The decomposition constant of *Cryptomeria* litter was the highest during the 34 months of incubation, while that of *Quercus* was the lowest at 16 months or later. The high decomposition rate in *Cryptomeria* is likely related to increased N content, while the low decomposition constant of *Quercus* might have originated from the change in chemical composition of the litter during its decomposition.

Table 1. Total N content (g·g^{-1}) and its ratio to the initial N content in decomposing litter of *Cryptomeria*, *Quercus* and *Pinus*. Litter bags were set on December 26, 2000. Values were means with standard errors in parentheses calculated from ten samples of a tree species in a site at a sampling date. Yamanaka et al. (2011) with permission

Site	Date (months of incubation)								
Litter species	Initial content	2002/4/18 (16)		2002/7/17 (19)		2003/7/15 (31)		2003/10/29 (34	
		content	ratio	content	ratio	content	ratio	content	ratio
Cryptomeria stand									
Cryptomeria	4.7	15.7	3.35	17.7	3.76	14.8	3.14	14.2	3.02
		(0.7)	(0.15)	(0.6)	(0.13)	(0.4)	(0.09)	(0.3)	(0.07)
Quercus	10.3	16.9	1.64	19.1	1.87	14.9	1.46	14.4	1.41
		(0.2)	(0.02)	(0.2)	(0.02)	(0.3)	(0.03)	(0.2)	(0.02)
Pinus	7.4	11.6	1.56	13.6	1.83	12.3	1.65	11.4	1.54
		(0.5)	(0.06)	(0.6)	(0.08)	(0.3)	(0.04)	(0.3)	(0.04)
Deciduous broad-leaved forest									
Cryptomeria	4.7	15.8	3.35	17.6	3.74	14.2	3.03	14.1	2.99
		(0.9)	(0.19)	(0.5)	(0.11)	(0.3)	(0.07)	(0.5)	(0.11)
Quercus	10.3	17.1	1.52	19.2	1.87	15.4	1.51	14.8	1.44
		(0.7)	(0.16)	(0.6)	(0.06)	(0.3)	(0.03)	(0.4)	(0.04)
Pinus	7.4	12.6	.70	14.5	1.95	12.1	1.64	11.9	1.60
		(0.3)	(0.04)	(0.5)	(0.07)	(0.1)	(0.02)	(0.3)	(0.03)

Salamanca et al. (1998) reported that the lignin content in the litter of *Q. serrata* strongly regulated the decomposition rate during the third year.

DLitter set in CS in the middle part of the slope decomposed faster than that in DBF at the upper site, while the water content of the litter was higher in CS than in DBF and the N content in the decomposing litter of the same species did not differ between the two sites (Table 3). The difference in decomposition rates between the sites is thus due to the water content of the litter. The ergosterol content in the litter in CS in the middle part exceeded that in DBF at the upper sites, although this difference was not statistically different at $P = 0.05$.

The present study showed that N_2-fixing activity in decomposing *Cryptomeria* litter was dependent on litter species, not on site. The N_2-fixing activity increased in both sites after 16 and 19 months of incubation respectively (Figure 1) could be explained by the change in litter quality during decomposition.

Table 2. Ergosterol content (µg g^{-1} dry matter) in decomposing litters of *Cryptomeria, Quercus and Pinus.* Litter bags were set on December 26, 2000. Values were means with standard errors in parentheses calculated from five samples of the same tree species in a site at a sampling date. Yamanaka et al. (2011) with permission

Site	Date (months of incubation)				
Litter species	2001/3/27 (3)	2002/4/18 (16)	2002/7/17 (19)	2003/7/15 (31)	2003/10/29 (34)
Cryptomeria stand					
Cryptomeria	108.0 (6.3)	400.4 (34.8)	358.3 (43.0)	120.4 (10.4)	54.4 (16.1)
Quercus	121.9 (11.6)	169.6 (15.6)	186.5 (58.1)	55.5 (8.5)	33.4 (9.0)
Pinus	51.0 (4.6)	249.1 (26.3)	325.8 (12.7)	127.8 (15.3)	52.6 (12.4)
Deciduous broad-leaved forest					
Cryptomeria	96.6 (9.0)	269.8 (21.70	257.9 (17.0)	115.2 (16.8)	95.2 (15.8)
Quercus	125.9 (4.1)	203.0 (18.5)	182.9 (34.2)	101.3 (8.5)	50.1 (14.4)
Pinus	45.4 (4.2)	166.7 (10.5)	376.0 (65.0)	123.3 (4.2)	75.7 (16.6)

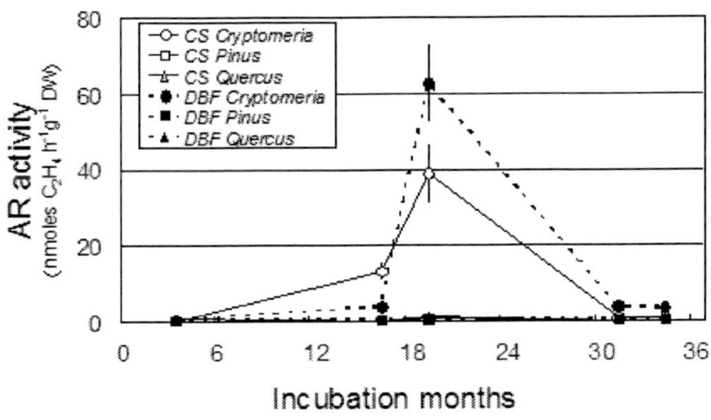

Figure 1. Acetylene reduction (AR) activity (nmoles C_2H_4 h^{-1} g^{-1} DW) in decomposing litters of *Cryptomeria*, *Quercus* and *Pinus*. Litterbags were set in *C. japonica* stand (CS) and deciduous broadleaved forest (DBF) on 26 December, 2000. Values were mean with standard errors calculated from ten samples of a tree species at a site on a specific sampling date. Yamanaka et al. (2011) with permission.

Table 3. *F*-values and sidnificance level for ANOVA of N content and its ratio to initial N content. Yamanaka et al. (2011)

Source of variation	*df*	N content		Ratio	
		F	*P*	F	*P*
Period of incubation	3	68.722	0.000	41.573	0.0000
Site	1	1.299	0.256	0.055	0.8153
Litter Species	2	165.444	0.000	1020.920	0.0000
Period x Site	3	0.273	0.845	0.133	0.9406
Period x Litter	6	3.402	0.003	3.189	0.0051
Site x Litter	2	1.206	0.301	0.913	0.4029
Period x Site x Litter	6	0.282	0.925	0.406	0.8745

The N dynamics in the *Cryptomeria* forest were also expected to advance differently from that in other forests during the decomposition of *Cryptomeria* litter. Estimation of the amount of N_2 fixation in the decomposition of *Cryptomeria* litter and its comparison with the amount of N immobilized by fungal colonization is necessary to characterize N dynamics in CS.

2.2. Greenhouse Gas Fluxes from Soil

Carbon dioxide (CO_2), methane (CH_4), and nitrous oxide (N_2O) are major greenhouse gases with contributions of 63, 18, and 6.1%, respectively (IPCC, 2007). In general, forest soils function as sources for CO_2 and N_2O and a sink for CH_4 (IPCC, 2007). Because these gases are produced and consumed by microorganisms and plants, soil properties such as temperature and water content influence the process (Hashimoto et al., 2011). Here, the characteristics of these gas dynamics are shown in relation to the slope position, i.e. on the bottom part, the lower slope in Japanese cedar trees (CS), and the upper slope in hardwoods (DBF).

The greenhouse gas fluxes were measured monthly from March 2004 to April 2006. The gas fluxes were measured by a closed-chamber technique according to the method of Sakata et al. (2004) and Morishita et al. (2007). Three cylindrical stainless steel chambers, 15 cm in height and 40 cm in diameter, were inserted in the soil to a depth of 3 to 5 cm at each site. A gas sample was taken into 30 ml glass bottle at 0, 10, 20, and 40 min after the chamber was closed. CO_2, CH_4, and N_2O concentrations in the bottle were determined by a gas chromatograph (GC) equipped with a thermal conductivity detector (TCD), a flame ionization detector (FID) and a electron capture detector (ECD), respectively. The fluxes of CO_2, CH_4 and N_2O were calculated from the change in the gas concentration in the chamber after closure.

The mean CO_2 flux on the upper slope (101 ± 70 mg C m^{-2} h^{-1}) exceeded that of other sites (bottom, 48 ± 34 mg C m^{-2} h^{-1}; lower, 65 ± 44 mg C m^{-2} h^{-1}) (Table 4). This trend was the same as the order of mean soil temperature, namely, the soil temperature on the upper slope was the highest among the sites (Table 4), while the CO_2 flux at each site showed a positive correlation with soil temperature (Figure 2). Relatively high CO_2 flux was observed on the upper slope compared to other parts, when data were compared under similar soil temperatures. Multiple regression analysis showed not only soil temperature but also soil water content influenced the CO_2 flux, but that the effect of soil water content was smaller than the soil temperature, according to the values of the standardized partial regression coefficient (Table 5). Therefore, the highest CO_2 flux on the upper slope was due to not only high soil temperature but also the soil water content in the sites. The annual CO_2 emission rate was estimated at 4.2 Mg C ha^{-1} at the bottom, 5.7 Mg C ha^{-1} on the lower slope, and 8.9 Mg C ha^{-1} on the upper slope, respectively. These values range from 3.06 to 10.57 Mg C ha^{-1} in Japanese forest soil (Isizhuka et al., 2006).

In the cedar site, the annual deposition rate of C through litterfall ranged from 1.1 to 1.4 Mg C ha^{-1} (Noguchi et al., 2009; section 3.2 in this chapter), while the ratio of CO_2 emission to litterfall in this site ranged from 4 to 5.

Table 4. Summary of greenhouse gas fluxes in the site

		Bottom			Lower slope			Upper slope		
	unit	mean	2004	2005	mean	2004	2005	mean	2004	2005
Temperature	°C	11.9	12.7	11.2	13.4	14.2	12.5	14.7	15.5	13.9
Water content	m^3m^3	0.67	0.70	0.64	0.24	0.23	0.25	0.15	0.15	0.15
CO_2 flux	Mg C m^{-2} h^{-1}	48	57	39	65	64	67	101	105	98
CH_4 flux	µg C m^{-2} h^{-1}	-7	-7	-6	-57	-58	-55	-65	-66	-64
N_2O flux	µg N m^{-2} h^{-1}	5.7	4.7	6.7	2.1	2.5	1.7	1.1	1.0	1.2

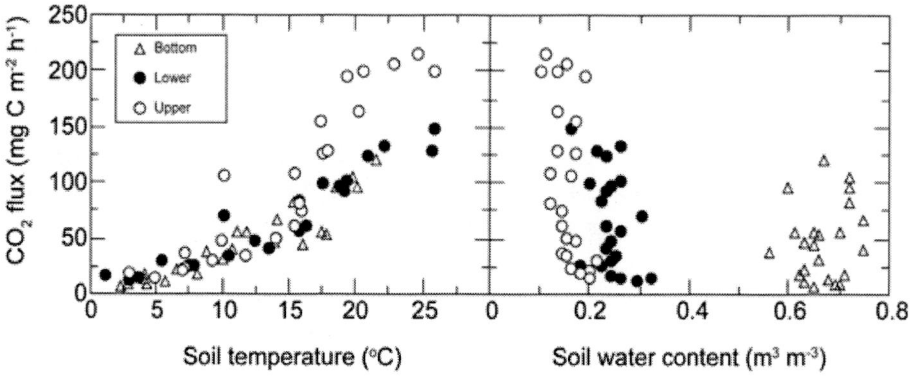

Figure 2. The relationship between CO_2 flux and soil temperature or moisture in the sites.

Table 5. The results of multi regression analysis in the sites

	standardized partial regression coefficient		R^2	*P*
	T	W		
CO_2 flux	0.82	-0.20	0.77	<0.001
CH_4 flux	-	0.96	0.93	<0.001
N_2O flux	0.39	0.45	0.27	<0.001

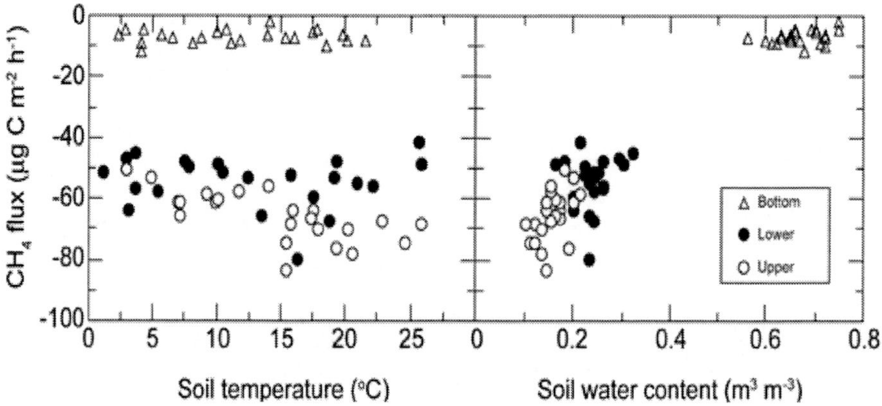

Figure 3. The relationship between CH_4 flux and soil temperature or moisture in the sites.

Raich et al. (1989) found C release from soil was nearly three times the C input from aboveground litter, and their data of coniferous forests in Japan ranged from 2.3 to 6.5. Before the CO_2 flux measurement, Japanese cedars were thinned and many residues such as litter, branch and stem had fallen and remained on the ground. These materials were thought to be a source of CO_2, but there was no clear increase in CO_2 flux.

The uptake of CH_4 was observed in each site in all measurements. The mean CH_4 uptake rates were as follows: on the upper slope (-65 ± 8 μg C m^{-2} h^{-1}) > on the lower slope (-57 ± 9 μg C m^{-2} h^{-1}) > at the bottom (-7 ± 2 μg C m^{-2} h^{-1}) (Table 4).

The mean values on the lower and upper slopes were similar to the reported value of -66 ± 20 μg C m^{-2} h^{-1} in Japanese forests (Morishita et al., 2007), while the value at the bottom was smaller than the value and similar to the value -3 ± 5 μg C m^{-2} h^{-1} in Japanese ash forest growing nearby a stream. The CH_4 flux was positively correlated with soil water content, but not with soil temperature (Figure 3), namely, the CH_4 uptake decreased with an increase in soil water content. According to multiple regression analysis, only soil moisture was selected as an explanatory value (Table 5). In fact, the soil water content was at the bottom was the highest among the site (Figure 3). After high rainfall, the bottom part was flooded several times during the summer season, CH_4 uptake activity was considered to have weakened due to the high soil water.

In contrast to the CH_4 uptake rate, the mean N_2O flux at the bottom (5.7 ± 8.6 μg N m^{-2} h^{-1}) was the highest among the site (Table 4). The values on the upper (1.1 ± 1.2 μg N m^{-2} h^{-1}) and lower slopes (2.1 ± 2.2 μg N m^{-2} h^{-1}) were similar to 1.88 ± 1.89 μg N m^{-2} h^{-1} as reported in Morishita et al. (2007). The N_2O flux was positively correlated with soil temperature and water content (Figure 4). According to the standardized partial regression coefficient of the multiple regression analysis, soil temperature and water content had the same effect on N_2O flux (Table 5). Furthermore, the high nitrification rate resulted in high N_2O flux on the lower and upper slopes of the site (Morishita et al., 2011). Therefore, high N_2O flux was due to the high soil temperature under high soil water conditions.

The annual N_2O emission rate was estimated at 0.50 kg N ha^{-1} at the bottom, 0.18 kg N ha^{-1} on the lower slope, and 0.10 kg N ha^{-1} on the upper slope, respectively. In contrast to C dynamics, the magnitude of N release as N_2O was very small compared with the litterfall rate of about 20 kg N ha^{-1} (Noguchi et al., 2009; section 3.2 in this chapter). In addition, the effect of the thinning was not clear in this study (Morishita et al., 2011). Therefore, the contribution of N_2O to N dynamics in this site was small whereas N_2O was one of the major greenhouse gases.

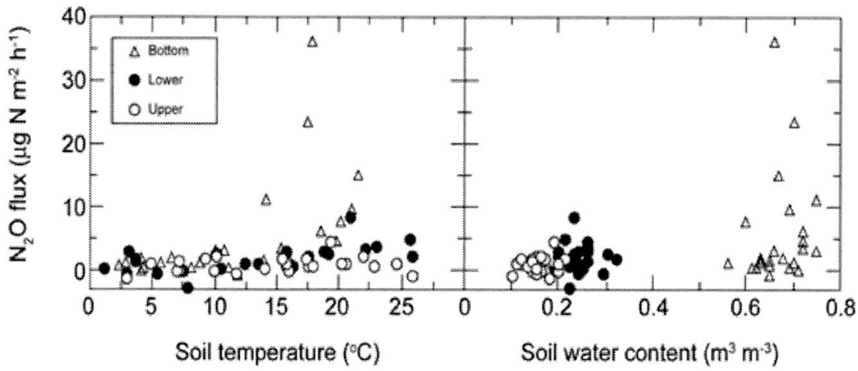

Figure 4. The relationship between N_2O flux and soil temperature or moisture in the sites.

2.3. Trace Metals Cycling

Human activities have resulted in the emission of huge amounts of trace metals, which has drastically altered the natural distribution of these metals and the biogeochemical cycles in terrestrial and aquatic ecosystems (Borg and Johansson, 1989; Nriagu, 1990). The metals are toxic to living organisms, even at low levels, and strongly influenced by human activities. Once emitted, trace metals can be transported long distances in air masses and contaminate terrestrial and aquatic environments. Lead and cadmium are emitted into the atmosphere from various anthropogenic sources: mining, smelters, various industrial processes, fossil fuel combustion, automobiles, and waste incinerators (Pacyna and Pacyna, 2001). Japan is one of the largest users of metals in the world (Asami, 2001), while trace metal emissions are also increasing in Asia (Pacyna and Pacyna, 2001).

Atmospheric deposition is a main pathway for lead and cadmium influx into forest ecosystems (Galloway et al., 1982). Thus, both the forest floor and mineral soil were impacted by the atmospheric deposition of lead and cadmium. Metals accumulated in forest ecosystems, however, are likely to be released into surface water if the deposition of trace element continues and the land use changes, and so on. The objective of this study was to evaluate the atmospheric lead and cadmium budgets and their biogeochemical cycling at KHC.

A) Lead and Cadmium Budgets

Bulk precipitation and throughfall samples were collected at KHC from April 2002 to May 2003. No significant anthropogenic lead and cadmium emission points were found in KHC. The annual lead and cadmium depositions in bulk precipitation at KHC were 8.9 and 0.8 g ha^{-1} yr^{-1}, respectively. The annual lead and cadmium depositions in throughfall at KHC were 5.3 and 0.4 g ha^{-1} yr^{-1}, respectively. Monthly lead and cadmium depositions in bulk precipitation increased in summer, due to the onset of the rainy season and the occurrence of typhoons (Figure 5). In addition, in the spring of 2002 and 2003, the yellow sand phenomenon was often observed, and Cd deposition in bulk precipitation increased simultaneously (Figure 5). This is considered attributable to transboundary air pollution.

In North America, Europe, and Japan, serious air pollution occurred during the 1960s to the early 1980s due to various anthropogenic emissions.

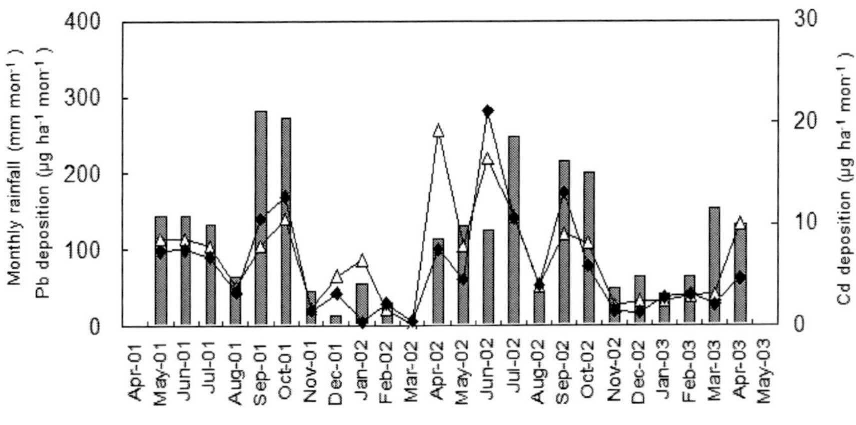

Figure 5. Temporal variation of lead and cadmium deposition in bulk precipitation at four forests. The column shows the monthly rainfall amount. Diamonds show lead deposition; left axes and triangles show cadmium deposition; right axes.

High concentrations of lead and cadmium in bulk precipitation at urban areas in Europe were reported 190 and 19 μg L^{-1}, respectively (Galloway et al., 1982). At a forest site in central Germany, lead and cadmium concentrations in bulk precipitation were reported at 38 and 1.2 μg L^{-1}, respectively, with resultant depositions of 405 and 13 g ha^{-1} yr^{-1} (Heinrichs and Mayer, 1977). These past concentrations and deposition rates were considerably high. Since the mid-1970s, stringent environmental controls have resulted in decreased emissions of trace metals from major point sources in North America, Europe, and Japan (Mukai et al., 1993). Due to these controls, the annual lead deposition at the Hubbard Brook Experimental Forest (USA) decreased drastically from 396 g ha^{-1} yr^{-1} in 1976 to 12.7 g ha^{-1} yr^{-1} in, 1989 (Johnson et al., 1995). The annual lead and cadmium depositions in the current study were similar to those recorded after 1990 at rural forests in North America and Europe, indicating obvious improvement in the atmospheric environment. A few recent studies have presented lead and cadmium concentrations in precipitation at urban and rural areas in Japan (Kobayashi, 2001; Tsumura and Yamasaki, 1998). In addition, Takeda et al. (2000) reported that the annual depositions of lead and cadmium at a rural site in Japan were 17.8 and 0.90 g ha^{-1} yr^{-1}, respectively. The concentration and input of lead and cadmium in these studies coincide with those found in the current study and the values were very similar. These results suggest that atmospheric lead and cadmium depositions are similar over a wide area in Japan, regardless of habitat type (forest, rural or urban area). Conversely, atmospheric lead and cadmium depositions at a remote site in Greenland were reported at 0.08 and 0.003 g ha^{-1} yr^{-1}, respectively (Sherrell et al., 2000). The lead and cadmium depositions in Japan were substantially greater (>3 orders of magnitude) than those at the remote sites (Figure 6). These differences indicate that the various anthropogenic emission sources of these metals remain widespread, with urbanization and an increase in automobiles and waste incinerators, despite reductions in emissions from large point sources.

Stream water chemistry is a consequence of various biogeochemical processes within watershed ecosystems, which consist of physical, chemical, and biological reactions. Many small catchment studies have been conducted to understand the processes having occurred at the watershed (e.g. at the Hubbard Brook Experimental Forest, and Solling forests). While atmospheric acid deposition affects the mobility of metals in soils and bedrocks, the leaching of toxic metals into drainage waters is one of the major concerns of current research (Frei et al., 2000). Moreover, lead and cadmium directly affect the health of aquatic creatures through bioaccumulation in the food web (Croteau et al., 2005). Elucidating the stream outputs of elements enables us to gain a more in-depth understanding of the various processes of elements cycling deeper understand, or assess the influence of toxic metals on living things in streams and downstream environments.

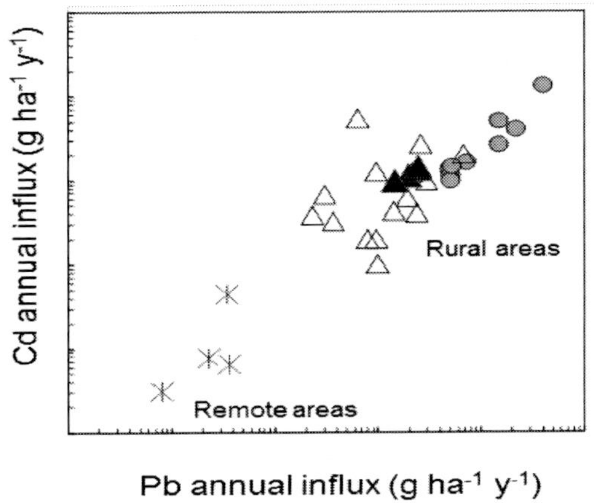

Figure 6. Comparison of lead and cadmium deposition in precipitation in rural forests and remote areas. Circles (○) show the lead and cadmium influxes in Japan (this study). Squares (◆) show lead and cadmium influxes for the period 1969-1990 (Heinrichs and Mayer, 1977; Jeffries and Snyder, 1981; Lindberg and Harriss, 1981; Durand et al., 1994). Triangles (▲) show the influxes of the period 1991-1999 (Aastrup et al., 1995; Avila and Rodrigo, 2003; Berg et al., 1994; Gélinas and Schmit, 1998; Lawlor and Tipping, 2003; Lawson, 2001; Ukonmaannaho et al., 2001). Stars (*) show the influxes in remote areas in 1993-1995 (Halstead et al., 2000).

The lead concentration in the stream water at KHC ranged from 3.4 to 84.8 ng L^{-1} and 1.2 to 27.4 ng L^{-1}, respectively (Figure 7). Although high lead concentrations were occasionally observed, there were obscure seasonal variations in the concentration for both elements in the stream water at KHC. Lead and cadmium concentrations in stream water were significantly low and steady during the base flow period. However, both elements increased by two to three orders during and right after rainstorms. In fact, heavy rain (>50mm day^{-1}) was reported at the site just before the collection of high concentration samples. During the storm flow, concentrations of lead and cadmium were attributed to increases in suspended sediment, linked to particle-bound metals (Duke et al., 1999). However, the concentrations of lead and cadmium showed low correlations with DOC at KHC. To accurately estimate the stream water outputs of these elements, the concentration must be determined at various flow periods. Comparing with the concentrations of lead and cadmium in rainfall in the Kanto district (Itoh and Yoshinaga, 2006), the concentrations of these metals were significantly low, approximately one sixty-fifth and one fifteenth of those of rainfall, respectively. This indicates that during the passage of elements through the forest ecosystem, most were retained within the same. However, acid deposition and land use change can accelerate the weathering rate of rocks, which may lead to leaching of lead and cadmium retained in the forest ecosystem. Thus, long term monitoring of the trace metals concentration in the stream water is key to assess the influence of human activities on the forest ecosystem.

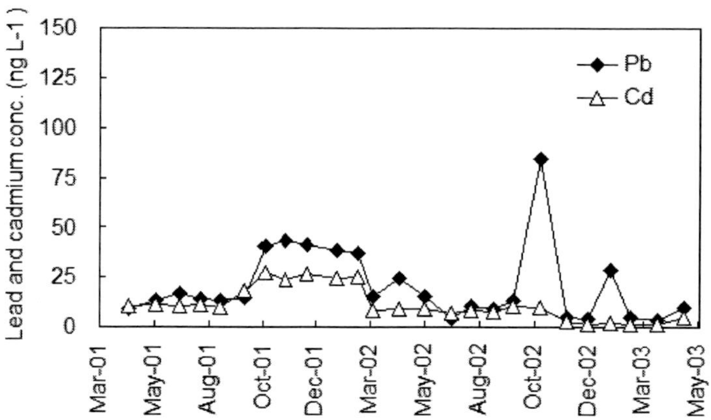

Figure 7. Temporal changes in concentrations of lead and cadmium in KHC stream water.

Atmospheric deposition is a most important pathway for lead and cadmium influx into forest ecosystems and the stream output is a main removal route for the export of their elements. A comparison of the concentration of elements between rainfall and stream water simply indicates the qualitative characteristics of the watersheds. When estimating the balance of these elements in a forest ecosystem, the small catchment study is a powerful tool and enables us to understand how biogeochemical processes exist. Several small catchment studies of trace metals have also been carried out in the northeastern USA and Northern and Central Europe (Van Hook et al., 1977; Stahr et al., 1980; Ukonmaanaho et al., 2001). In these studies, the input of lead and cadmium were significantly smaller than the output, due to physical, chemical, or biological processes for each element within the watershed. However, previous studies were conducted in cooler temperature areas than Japan; thus, various biogeochemical reaction rates would differ from those of the forest ecosystems in Japan.

The annual mean lead input was estimated to be 23 g $ha^{-1}y^{-1}$ and its output was approximately 0.8 g $ha^{-1}y^{-1}$ (Figure 8). The annual mean cadmium input was estimated to be 1.5 g $ha^{-1}y^{-1}$ and the output was approximately 0.2 g ha^{-}

$^{1}y^{-1}$ (Figure 8). Lead retention was very high (>96%), as was cadmium retention (86%) at KHC.

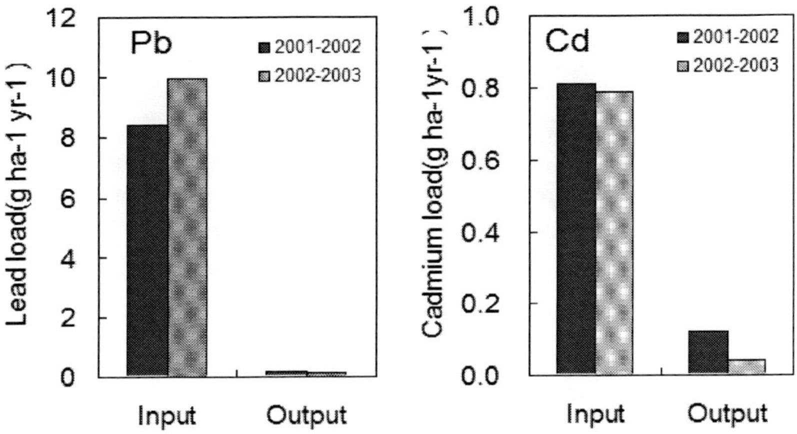

Figure 8. Annual input-output budgets of lead and cadmium at the KHC watershed.

Input–output budgets indicated that most of the atmospheric inputs of lead and cadmium were retained somewhere within this catchment. Smith and Siccama (1981) reported that lead retention was 97 % at HBEF in the USA. Lindberg and Turner (1988) showed lead and cadmium retentions at the Walker Branch, ranging from 97 to 99 % and 74 to 92 %, respectively. Based on recently reported values, the lead retention ranged from 94 to 97 % with cadmium retention of 75 % at boreal forests (Ukonmaanaho et al., 2001). The retention rates of this site are similar (Lead retention rate: >96%, cadmium: 86%) to those reported for cool temperate and boreal forest ecosystems. High lead and cadmium accumulated in Japanese forests was also reported in cool temperate and boreal forests. The retention of these trace elements in forest ecosystems seems to be responsible for adsorption in the canopy and soil, plant uptake, and/or accumulation in biomass. However, based on input-output budgets, it is not possible to determine the place and amount for lead and cadmium retention. Several forest ecosystems showed similar input-output ratios for lead and cadmium, regardless of differences in climate and site environments, while the same processes for the dynamics of these metals would work within forest ecosystems. However, the mechanisms have not yet been revealed for this forest ecosystem.

Inputs of lead, cadmium, and inorganic-N exceeded outputs of these elements, as did inputs of sulfur and calcium. The major element retentions calculated were as follows: 68 % for inorganic-N, -110 % for sulfur and -298 % for calcium respectively. The positive budget of nitrate would result from not only active plants uptake but also several complex processes related to hydrology, immobilization, denitrification, to which N fixation may contribute (Hornbeck et al., 1997; Campbell et al., 2004). The negative budgets, conversely, result from leaching from weathering rock or weak adsorption to forest soil (Likens and Bormann, 1995; Watmough and Dillon. 2003a). The high accumulation of lead and cadmium demonstrated in the forest ecosystem suggested the strong affinity of these elements to organic matter or clay minerals (Stahr et al., 1980; Kabata-Pendias, 2001). Consequently, most of the atmospheric inputs of lead and cadmium were also retained within the forest ecosystem in Japan.

B) Biogeochemistry of Lead and Cadmium

The annual input-output budgets of lead and cadmium in the Japanese forest indicated that most atmospheric inputs of lead and cadmium were retained in forest ecosystems. However, since it is difficult to determine the place and amount of lead and cadmium retained completely, study of the distributions of these metals is indispensable to evaluate their dynamics and the cycling of trace elements in the forest ecosystem.

The cycling of trace elements in forest ecosystems has been studied in North America and Europe since the 1970s (Heinrich and Mayer, 1977; Van Hook et al., 1977; Likens et al., 1977: Turner et al., 1985). These studies generally concluded that the forest floor was a large sink for trace metals due to their tight affinity to organic matter. Therefore, the dynamics of the organic matter may impact the behavior of the retained metals. Since the study areas were cool temperate and boreal climate regions with relatively slow decomposition rates, trace element cycling in temperature-moist forests in Japan may differ because the decomposition rate in Japan is faster than those of North America and Europe. The input rates of the metals also influence their dynamics in the forest floor. Friedland et al. (1992) demonstrated that a recent reduction of anthropogenic lead emissions in North America resulted in a decrease in lead concentration in the forest floor and the migration of lead to forest soil in the northeastern USA. Thus, the determination of cycling and detailed distributions of toxic trace elements, such as lead and cadmium, in all forest ecosystem components is important to elucidate the elements cycling in the ecosystem. These studies deal with the geochemical elements flow, which mainly concerns aquatic pathway; precipitation, soil solutions, ground water, and stream water. Forest ecosystems have other channels through plants, although the elements are actually transferred by water. The cycling of essential major elements (e.g. N, P, and K) by plants has been investigated in many forest ecosystems (Duvingneaud and Denaeyer de Smet, 1970; Likens et al., 1977; Staff and Berg, 1977; Whittaker et al., 1979; Sollins et al., 1980). However, the cycling of toxic trace elements through plants has been rarely studied in forest ecosystems. Lead and cadmium are presumed to have small active-uptakes into biomass, except for special species; hyper-accumulators. In boreal and cool temperate forests, the uptake of lead and cadmium by vegetation is considered negligible, compared to the atmospheric influxes (Berthelsen et al., 1995; Starr et al., 1998; Bindler et al., 1999; Reimann et al., 2001). However, such studies have not been yet conducted in Japan, under a temperate moist climate.

The mean and standard deviations of lead and cadmium concentrations in Japanese cedar and broad leaved trees in the watershed are shown in Figure 9. All parts of the vegetation contained lead and cadmium, with no significant differences between the tree species of lead and cadmium concentration. The lead and cadmium concentrations peaked in the fine roots for every tree species.

There were also relatively significant differences among the parts of vegetation, with peak concentrations of lead and cadmium in each tree species approximately 10 to 50 times larger than the minimum levels. The fine roots function as entrances, accepting various elements from the mineral soils, in which lead and cadmium have accumulated in bulk; hence their concentrations exceed those elsewhere. Conversely, the high concentration of fine roots might be overestimated due to the incomplete separation between samples and soil particles. Lead and cadmium were highly concentrated in the H layer of the A_0 layer (O layer) at both stands, while the trace metal concentration of the A_0 layer increased with the decomposition of litter (Berg and McClaugherty, 2004). This result indicates that it was difficult to mineralize lead and cadmium and transfer them into the A_0 layer. The concentrations of lead and cadmium in litterfall at the Japanese cedar stand ranged from 0.24 to 5.10 mg kg^{-1} and 0.00 to 0.25 mg kg^{-1}, respectively. In litterfall at the broad leaved stand meanwhile, the concentrations of lead and cadmium ranged from 0.74 to 10.94 mg kg^{-1} and from 0.05 to 0.99 mg kg^{-1}, respectively. There were significant differences among the parts of litterfall samples as well as the vegetation in the watershed. Lead concentrations in the forest soil ranged from 6.80 to 40.3 mg kg^{-1}, while the cadmium concentration in the forest soil ranged from 0.05 to 0.74 mg kg^{-1}. The concentrations of lead and cadmium were the highest in the surface soil horizon and decreased with soil depth in most of the forest profiles. These patterns of lead and cadmium distribution were consistent with those in many previous studies (e.g. Stahr et al., 1980; Heinrichs and Mayer 1977; Friedland and Johnson 1985).

The pools of lead at the Japanese cedar and broad leaved stands were the largest in the stem wood, but, the largest cadmium pools differed from the tree species. In the Japanese cedar stand, the cadmium pool was the largest in the stem wood, but in the broad leaved stand, it was observed in the fine roots, due to the different amounts of biomass between the two stands.

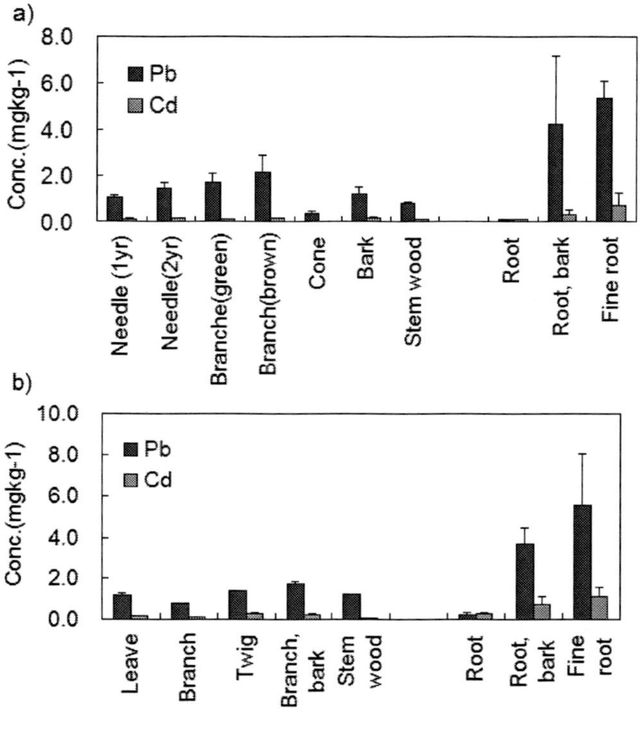

Figure 9. Lead and cadmium concentrations in some vegetation components. a) Japanese cedar, b) Broad leaved trees. Root: ≥2 mm in diameter, fine root: <2 mm in diameter.

Total pools of lead and cadmium in the above ground vegetation were 172.8 g Pb ha^{-1} and 20.4 g Cd ha^{-1} in the Japanese cedar stand, and 65.0 g Pb ha^{-1} and 5.7 g Cd ha^{-1} in the broad leaved stand, respectively. Those in the vegetation below ground were 23.2 g Pb ha^{-1} and 6.3 g Cd ha^{-1} in the Japanese cedar stand, and 30.1 g Pb ha^{-1} and 7.1 g Cd ha^{-1} in the broad leaved stands respectively. Generally, the pools of lead and cadmium in the aboveground vegetation exceeded those below ground, except for the cadmium pools in the broad leaved stand below ground. In the A_0 layer, the metal contents were estimated at 104.7 g Pb ha^{-1} and 4.9 g Cd ha^{-1} in the Japanese cedar stand, and 98.3 g Pb ha^{-1} and 9.6 g Cd ha^{-1} in the broad leaved stand, respectively. The pools of lead and cadmium of the five profiles ranged from 39.9 to 142.7 kg Pb ha^{-1} and 0.5 to 1.0 kg Cd ha^{-1} in the watershed. Both element pools varied widely among the profiles in the watershed due to the variation in soil bulk density.

The biomass increments in coniferous and broad leaved stands were 5.4 and 3.0 tons ha^{-1} yr^{-1}, respectively, including root biomass increment in both stands. In the KHC meanwhile, the vegetation uptake of lead was 32 g ha^{-1}yr^{-1} at the Japanese cedar stand and 13 g ha^{-1}yr^{-1} at the broad leaved stand, respectively, while the uptake of cadmium was 2.0 g ha^{-1}yr^{-1} at the Japanese cedar stand and 1.7 g ha^{-1}yr^{-1} at the broad leaved stand, respectively.

The annual release by litterfall was calculated based on the lead and cadmium concentrations and amounts of litterfall. In the KHC, the lead release by litterfall was 28.4 g ha^{-1}yr^{-1} at the Japanese cedar stand and 8.2 g ha^{-1}yr^{-1} at the broad leaved stand, respectively. The release of cadmium was 1.3 g ha^{-1}yr^{-1} at both the stands. The annual vegetation uptakes of lead and cadmium exceeded their release by litterfall in both stands, although these balances between uptake and release may not lead to an increase in the amounts of lead and cadmium in the A_0 layer.

At KHC, the distributions of the lead and cadmium concentrations showed similar patterns in both the Japanese cedar and broad leaved stands. The forest soil, A_0 layer, and vegetation were the major pools of lead and cadmium in the forest ecosystem, with the concentration peaking in the surface soil layer, while mineral soil showed the largest pools for both elements. The forest ecosystem functions as a sink for lead and cadmium and lead in particular was accumulated and homogenized in the organic-rich fraction of the soil for an extended period (Friedland et al., 1985). However, Miller and Friedland (1994) reported that accumulated lead and cadmium were released more rapidly from forest catchments than expected due to the environment changes. Global warming, acidification of the environment and land-use changes may alter the dynamics of these toxic metals in forest ecosystems and the accumulated metals in forest ecosystems may become a source of these metals.

The annual increments of lead and cadmium in vegetation are considered fluxes, which enter the biomass pool either via root uptake or atmospheric deposition, but the proportions of their contributions cannot be divided in this study. Lead and cadmium are not essential elements and not actively consumed by plants. However, if the elements were in soluble form in the A_0 layer and mineral soil, plants could take them up with essential elements. It is therefore suggested that the rate at which these elements flow within the forest ecosystem is controlled by the activities of plants and soil microorganisms. In the KHC forest ecosystem, the uptake of lead and cadmium by vegetation (mostly roots) exceeded those of the cool temperate and boreal forests. Compared to the annual inputs of lead and cadmium, the amounts of cycling through vegetation (internal cycling) were not negligible orders in KHC. The vegetation plays an active role in the retention of lead and cadmium in this forest ecosystem. Conversely, the amount of lead in the forest floor was smaller than those in boreal and cool temperate forests. In the forest ecosystem, internal cycling is one of the functions of prevention against the migration of these toxic trace metals. It is suggested that these results from the rate of organic matter decomposition and vegetation intake, since the decomposition rate of organic matter and plant activity increases with rising temperature and moisture.

Concerning the vegetation components, the concentrations of both metals peaked in the fine roots, and were particularly high in the surface soil layer. Approximately 95-99 % of the total amounts of lead and cadmium in this forest ecosystem were in the forest soils, whereas the accumulations of lead and cadmium in the A_0 layer were limited in this forest ecosystem due to the rapid litter decomposition and the effective internal cycling through vegetation. The internal cycling, which consisted of plant root uptake and litterfall release, played a significant role in the cycling of lead and cadmium in this forest ecosystem. Although the A_0 layer was thin in the temperate moist forest, the internal cycling functioned effectively as a reservoir.

It has not yet been accurately elucidated where atmospherically derived lead and cadmium are distributed in the forest ecosystem, while there is also a need to understand the stability or mobility of the accumulated toxic metals in the forest ecosystem in future.

Atmospheric deposition is the major pathway of lead input into forest ecosystems, and most of the deposits are derived from anthropogenic emission sources (Nriagu, 1989; Sirahata et al., 1980). However, the lead output from forest ecosystems is considerably lower than the input, which occurs through stream discharge. Consequently, most atmospheric-derived lead is retained within vegetation, the forest floor, and the soil (Heinrichs and Mayer, 1980; Smith and Siccama, 1981; Friedland and Johnson, 1985; Miller and Friedland, 1994). Among these components, significant accumulations of lead are often observed in the surface soil layer and forest floor, because lead has a strong affinity to organic matter (Adriano, 2001), which results in a long residence time. For example, in North America and Europe, the residence times in the forest floor were estimated as 17 to 500 years (Miller and Friedland, 1994; Bindler et al., 1999; Kaste et al., 2003).

In nature, lead has four stable isotopes: ^{204}Pb, ^{206}Pb, ^{207}Pb, and ^{208}Pb. Since the lead isotopic compositions are distinct between anthropogenic lead in airborne particulate matter and geogenic lead in deep soil profiles (Hansmann and Köppel, 2000), several studies have used these differences to determine the origins of lead accumulated in forest soils and peat bogs (Bindler et al., 1999; Nowack et al., 2001; Hernandez et al., 2003; Haack et al., 2003; DesJardins et al., 2004). The isotope method confirmed that large quantities of lead in the surface soil layers and forest floor originated in airborne particles. Watmough and Hutchinson (2004) reported that approximately 60% of lead in the surface soil, 75% in the forest floor, and 65 to 100% in vegetation was derived from anthropogenic pollutants, in rural woodland in Ontario, Canada. This means that the lead accumulated in the surface soil had different sources, not only airborne particles but also geogenic origins, which suggests that biogeochemical cycling is important for understanding lead dynamics in forest ecosystems. Accordingly, different ecosystems would have different lead cycling characteristics owing to predominant vegetation, soil type, and geology. For example, in North American and European forests, where several studies have been conducted, a thick forest floor functions as a lead reservoir in the ecosystem (Heinrichs and Mayer, 1980; Friedland and Johnson, 1985; Turner et al., 1985; Miller and Friedland, 1994). In temperate Japanese forests, however, the thinner forest floor limits the lead accumulation (Itoh et al., 2005); hence the behavior of atmospheric-derived lead in Japanese forest ecosystems might differ from that in North America and Europe.

The relative proportions of atmospherically and geologically derived lead in the forest ecosystem components were calculated using the end member method (Shotyk et al., 1997; Bindler et al., 1999). A simple binary mixing model was used:

Atmospherically-derived Pb (%)

$$= \frac{^{206}Pb/^{207}Pb\ sample - {}^{206}Pb/^{207}Pb\ geogenic}{^{206}Pb/^{207}Pb\ atmospheric - {}^{206}Pb/^{207}Pb\ geogenic} \times 100 \quad (1)$$

where the $^{206}Pb/^{207}Pb$ sample is the $^{206}Pb/^{207}Pb$ ratio of the sample in the forest ecosystem, $^{206}Pb/^{207}Pb$ geogenic is the $^{206}Pb/^{207}Pb$ ratio of the geogenic lead, and $^{206}Pb/^{207}Pb$ atmospheric is the $^{206}Pb/^{207}Pb$ ratio of the atmospheric lead. For geogenic lead, the mean $^{206}Pb/^{207}Pb$ ratios of the subsoil (20–130 cm) were used, because volcanic deposits on the bedrock constitute the predominant parent material of the soil in the stand. For atmospheric lead, the mean $^{206}Pb/^{207}Pb$ ratios of bulk precipitation were used. From 96 to 99% of total lead in bulk precipitation is derived from anthropogenic origins (Sirahata et al., 1980; Nriagu, 1989).

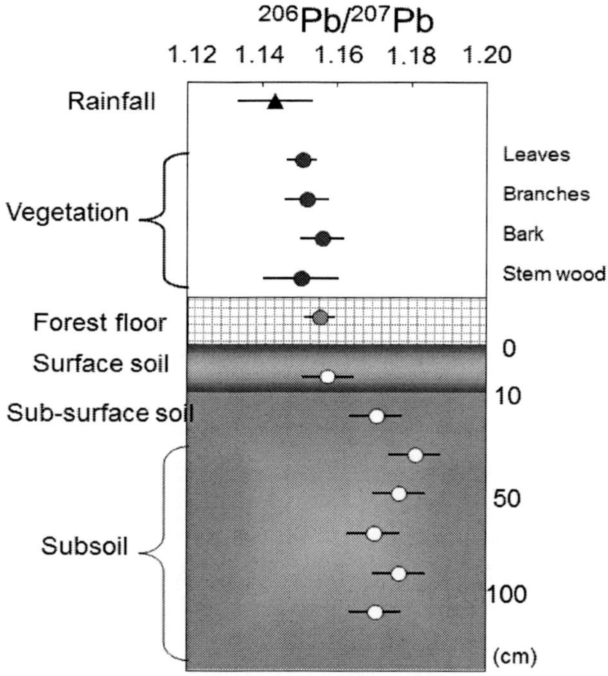

Figure 10. Vertical distribution of the lead isotope ratios ($^{206}Pb/^{207}Pb$) in the Japanese cedar ecosystem. Triangles and circles show the mean lead isotope ratios ($^{206}Pb/^{207}Pb$). Lateral bars show the standard deviation of lead isotope ratios ($^{206}Pb/^{207}Pb$) in each component.

The $^{206}Pb/^{207}Pb$ isotope ratios in the vegetation, forest floor, and surface soil showed intermediate values between those of anthropogenic and geogenic lead through mixing of these sources by biogeochemical cycling in the KHC ecosystem (Figure 10: Itoh et al., 2007). The incorporation of atmospherically derived lead into vegetation has two possible pathways: direct incorporation in the leaves and bark, and uptake from roots.

As Watmough and Hutchinson (2003b) reported, lead incorporation from bark is not likely to be a major pathway to wood, while we also consider that the direct incorporation of lead from bark and needles does not contribute extensively to lead cycling, because $^{206}Pb/^{207}Pb$ ratios in bulk precipitation were significantly lower than those of bark and needles. Lead would be absorbed by roots mostly from the mineral soil, because the forest floor is thin and no fine roots exist there. Some lead in vegetation is returned to the surface soil by litterfall, which establishes the biological cycling of lead in the forest ecosystem. Thus, vegetational factors such as the composition of vegetation, acquisition of lead, root distribution, and litter production are important for determining lead cycling in forest ecosystems. According to the simple binary mixing model (Eq. 1), 55% of lead in the surface soil (0–10 cm) was derived from the atmosphere, on average. In the sub-surface soil layer (10–20 cm), meanwhile, the atmospheric contribution decreased rapidly to 16%. Sixty-two percent of lead in the forest floor originated from anthropogenic sources. In the vegetation, the atmospheric origin was 71% on average, although the percentages among plant components differed slightly.

The lead in the forest floor is likely to migrate rapidly to the surface soil layer due to quick litter decomposition. The lead isotope ratios in the mineral soils suggest that lead from the decomposing forest floor permeates only into the surface soil (0–10 cm) (Figure 10), probably due to the high affinity of lead for soil organic matter. The high carbon content (120.7 mg g^{-1}) in the surface soil, which seems to be influenced by volcanic ash, may prevent lead from migrating downward a long distance, while organic matter and mineralogical characteristics in the soil would influence the lead dynamics of the ecosystem.

3. Effects of Thinning on Material Transformation

3.1. Nitrogen Supply to the Soil with Decomposing Green Needles Returned by Non-Commercial Thinning

In non-commercial thinning, no stem wood was harvested from the forest and it existed in the site for an extended period. The amounts of N storage in stem wood remained constant for extended years (Inagaki et al., 2009) and we believe the majority of the litter returned by non-commercial thinning was not in contact with the forest floor. Because slash containing leaved branches was not cut down from stem wood, this may affect the nutrient supply to the soil, with decomposing litter comparing ordinary thinning with stem wood harvesting. Accordingly, we discussed N dynamics with decomposing litter after thinning with a focus on the site difference of litter existence in this study.

A litter bag experiment was conducted in the CS. Litterbags containing 10g of dry weight green needles were installed on the forest floor and in the air (Figure 11) on April 2004 at three plots (5 x 5 m). The mesh size of the litterbag was 1 x 1 mm. In each plot, sampling was subsequently conducted until two years after and five bags had been collected during each sampling period. The collected needle samples were separated into < 2 mm and >2 mm using a sieve and oven dried at 70°C until the weight remained constant. These samples were weighted and fine ground >2 mm samples for N and C contents determined by an NC analyzer (Sumigraph NC-900).

Figure 11. Litter bag installation on the forest floor (left) and in the air (right).

The decomposition rate was also calculated by deducting the installed initial weight and sample weight of >2 mm in the litterbags during each sampling period.

We estimated that the amount of N returned by thinning was 84 kgN ha^{-1} and 67 % of them existed in the air without contact with the floor (Hirai et al., 2011). This means that 59.9 ± 43.2 kgN ha^{-1} existed in the air and 24.5 ± 11.3 kgN ha^{-1} existed directly on the forest floor.

The decomposition rate did not differ in the initial year between treatments but were statistically higher in the ground installed needles than those of the air after one year (Table 6). The values were 0.78 ± 0.04 in the ground and 0.58 ± 0.04. The water contents in the collected needles were also higher in the ground needles than the air needles for every sampling (Table 7).

The nitrogen concentrations in the remaining green needles increased from the initial 11.5 ± 0.5 to 23.4 ± 1.5 g kg^{-1} for ground installed needles and 23.6 ± 1.4 g kg^{-1} for air installed needles (Table 8). The installed air for needles was stable after 6 months but that of those installed on the ground increased continuously for two years. The N contents after a one year experiment was higher in ground installed needles 23.4 ± 1.5 g kg^{-1} than those in air needles 20.1 ± 0.6 g kg^{-1}.

The mass loss of N in the bag was calculated during the experiment (Table 9). It was observed that the N mass in the needles (needle mass x N content) increased in the first year and decreased in the second year decomposition in both treatments. The peak nitrogen mass decrement started quarter after in air and six months after in ground install treatment.

Table 6. Decomposition rate in green needle litters of Japanese cedar

Site	Date (days of decomposition length)			
	2004/7/28 (82)	2004/11/2 (179)	2005/5/13 (371)	2006/4/14 (707)
Air	0.247 (0.013)	0.419 (0.008)	0.427 (0.012)	0.575 (0.041)
Ground	0.233 (0.030)	0.415 (0.034)	0.458 (0.024)	0.783 (0.039)

Litterbags were set on 7 May 2004, Values were mean with standard deviation in parentheses in a site, n=5.

After one year of decomposition, an average of 99 percent of the initial N mass existed in the air installed treatments and 109 percent in the ground installed treatment in the litter bag.

No difference in decomposition rates during the first year was observed between treatments (Table 6). The lack of difference in the substrate quality of needle litter was considered attributable to the fact that needle samples for the experiment were collected in this site, and water soluble elements in the form of sugars and/or low molecular compounds were decomposed selectively at an early stage (Ono et al., 2009). After one year decomposition, decomposition was higher in ground installed needles than air needless. As cellulose and hemi-cellulose decomposed after one year, ground installed needles, which were easily attached from soil animals and microbes, decomposed more rapidly than air installed needles. It was also affected based on differences in the water condition in needles installed at different sites (Table 7).

Nitrogen immobilization by soil microbes occurred during decomposition. The difference in the N content increment of decomposing needles was affected with and without contact with the ground. For these reasons, the N contents of the ground installed needles increased incrementally faster than those of air (Table 8), while a nitrogen content increment was also observed in the air installed needles in this study. Nitrogen fixation activity appeared in decomposing litter, particularly in the F1 horizon (Nioh and Haruta, 1986). In our study, the effect of N immobilization with soil microbes did not take air installed needles into account, because these did not come into contact with the ground. This means that N_2 fixation occurred on Japanese cedar needles, even without contact with the soil, during the early-stage decomposition process.

The mass loss of litter progressed rapidly (Table 6) with increasing N contents (Table 8). The rate of N content increment also exceeded the mass loss rate in the first quarter, while the nitrogen mass in the needles after decomposition increased regardless of treatments during this period (Table 9). The nitrogen mass peaked three months later in air installed needles and 6 months later in ground installed needles during decomposition, whereupon the N release started.

The nitrogen mass of fresh cedar leaves felled by pruning reportedly increased until six months after decomposition (Aiba et al., 1983). This result supports our study, showing that nitrogen mass exists at 120 to 140 percent of initial mass at six months after decomposition (Aiba et al., 1983). In our study, an average of 120 percent of the initial mass in ground installed needles was almost the same, but the average of 100 percent in air installed needles was low. Our results indicated that the length and mass of N immobilization and/or fixation during the decomposition process differed depending on the installed site. We clarified that the N mass increment was short and low in air installed needles compared to those on the ground.

The green needle mass of Japanese cedar existing without contact with the ground returned by non-commercial thinning conducted in this watershed exceeds that of those with contact. We calculated the supplied N mass during decomposition in the watershed level. The nitrogen supply to the soil started from six months to one year in the site without contact and one to two years in the site with contact (Figure 12). Although these starting times were late, based on three to six months comparing the release of immobilized and/or fixed N_2, the results were the same as when observing changes in N mass loss, namely swiftly in air installed and late in ground installed needles (Table 9). During this period, the amounts of supplied N were small and considerable N existed compared to the initial period in the ground contact site. The nitrogen supply started one year after decomposition in the watershed level.

Table 7. Water contents(%, w/w) in green needle litters of Japanese cedar

Site	Date (days of decomposition length)			
	2004/7/28 (82)	2004/11/2 (179)	2005/5/13 (371)	2006/4/14 (707)
Air	21.7 (4.1)	72.0 (3.9)	18.3 (1.4)	61.9 (17.6)
Ground	41.6 (2.0)	78.3 (1.0)	53.8 (1.6)	72.3 (1.1)

Litterbags were set on 7 May 2004, Values were mean with standard deviation in parentheses in a site, n=5.

Table 8. Changes of N contents (g kg^{-1}) during decomposition in green needle litters of Japanese cedar

Site	Date (days of decomposition length)				
	Initial (0)	2004/7/28 (82)	2004/11/2 (179)	2005/5/13 (371)	2006/4/14 (707)
Air	11.5 (0.5)	16.6 (0.3)	20.9 (0.6)	20.1 (0.6)	23.6 (1.4)
Ground		17.5 (1.4)	23.5 (2.2)	23.4 (1.5)	22.3 (1.2)

Litterbags were set on 7 May 2004, Values were mean with standard deviation in parentheses in a site, n=5.

Table 9. N mass loss during decomposition in green needle litters of Japanese cedar

Site	Date (days of decomposition length)				
	Initial (0)	2004/7/28 (82)	2004/11/2 (179)	2005/5/13 (371)	2006/4/14 (707)
Air	115.7 (4.8)	125.1 (6.0)	120.8 (3.7)	114.6 (3.7)	100.0 (12.1)
Ground		134.1 (1.6)	136.8 (3.3)	125.6 (8.3)	48.9 (10.8)

Litterbags were set on 7 May 2004, values were mean with standard deviation in parentheses in a site, n=5, Nitrogen mass (mgN bag^{-1}) = needle mass x nitrogen content.

Between one and two years after, the N supply increased, up to 7.9 ± 6.3 kgN ha^{-1} in the site without contact and 14.1 kgN ha^{-1} in the site with contact respectively. As the increment of the decomposition rate between first to second year was high, 64 percent of the total supplied to the soil was contributed from within the ground contact site. The amounts of N supplied to the soil over the two years of decomposition were 22 kgN ha^{-1}, which means 26 percent of the N returned by thinning was supplied by the decomposition process during this period.

The amounts of N returned by thinning were 84 kg ha^{-1}, with a thinning rate of 33 percent of trees ha^{-1} in our site. This value is higher comparing Harada et al. (1972), citing an N mass stock in foliage of 120.0 kgN ha^{-1} in 38 year-old Japanese cedar stand in moderately moist brown forest soil (Forest Soil Division, 1972). In this study site, N was supplied at 33-45 kgN ha^{-1} yr^{-1} by litter fall (See section 3.2 in this chapter), whereupon N amounts in green needles returned by thinning were two times higher than those returned by annual litter fall. The amounts of N supply decreased 20-24 kgN ha^{-1} after thinning (See section 3.2 in this chapter), and the returned N mass by thinning to that directly at the ground surface was 25 kgN ha^{-1} (Hirai et al., 2011). These values are equivalent to the difference between the litter fall input of thinning and non-thinning stands. The litter fall input after thinning decreased due to a drop in standing tree numbers. Thinning supplies higher amounts of N at a time than normal litter fall. However, the N amounts directly supplied to the ground surface were equal as decremental amounts of litter were input before and after thinning respectively. The effect of slowing the N supply to the soil involved an important process in which the majority of the N mass returned by thinning existed in the air and was not in contact with the surface.

Our results indicate that N in green needles returned by thinning was not released in the initial year of decomposition, since N immobilization and/or fixation dominated when decomposition was compared (Tables 6 and 8). This means that green needles returned by thinning have a function that stores the N temporarily and releases it later. 67 percent of returned N by thinning existed in the air site which was not in contact with the ground, and in which the N supply to the soil started faster than in existing ground needles (Figure 12). Conversely, the existing N mass was small and while the commencement of the N supply to the soil in the ground site was delayed, significant N was supplied between one and two years after decomposition. Accordingly, the N mass distribution of green needles returned by thinning differed when the air and ground sites were compared, as did the decomposition rates and N supply patterns for both. These results suggest it may be possible to adjust the N supply by managing litter treatments.

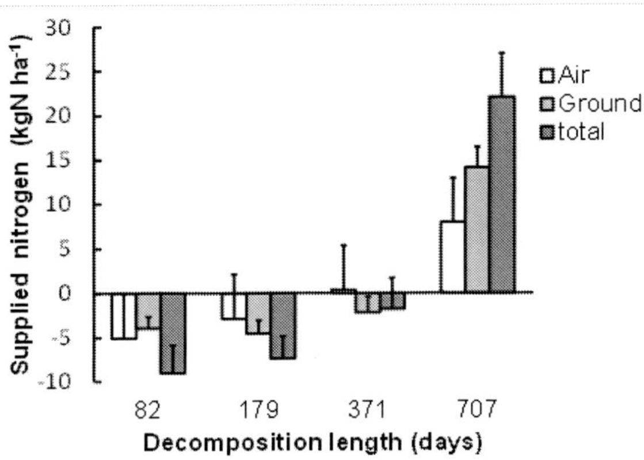

Figure 12. Amounts of supplied nitrogen to the soil from decomposing green needles of *C. japonica*.

3.2. Dynamics of the Aboveground C and N

Forest ecosystems have various ecosystem functions, a key example of which is C assimilation. Recent reports suggested that accumulated C in tropical, temperate and boreal forests amounted to 71-146, 61-155 and 8-24 MgC ha^{-1}, respectively (reviewed by Keith et al., 2009). Nowadays, this function of forest ecosystems is expected to mitigate the global warming likely to result from increasing atmospheric CO_2. Another important function of forest ecosystems is the uptake of nutrients, which can mitigate the eutrophication of stream water. In particular, N uptake by forest trees is important because the deposition rate of N has been increasing due to human activities such as the use of fossil fuels and industrial N_2 fixation (Galloway et al., 2003). Therefore, there is a need to deepen our understanding of C and N dynamics to evaluate the ecosystem functions of forest ecosystems. Although a number of studies have been reported for C and N dynamics in forest ecosystems, only limited information exists on the effects of forest management practices. Therefore, in this study, we aimed to elucidate the effects of thinning practice on C and N dynamics in a *C. japonica* plantation.

To estimate the aboveground biomass and aboveground C and N contents of *C. japonica* trees, allometric equations were established using six sample trees of different sizes (Table 10). The diameter at breast height (DBH) and the aboveground biomass of the sample trees ranged from 4.8 to 24.8 cm and 4.2 to 201.1 kg, respectively, while the Carbon and N concentrations in the aboveground organs of the sample trees were determined using an NC analyzer (Sumigraph N22F).

Table 10. Allometric equations for estimating aboveground biomass and aboveground carbon and nitrogen storage in *C. japonica* trees in Katsura experimental watershed

	a	b	R^2	*P* Value
Biomass (kg) = a DBH^b	0.0974	2.39	0.99	<0.0001
C content (kg) = a DBH^b	0.0450	2.40	0.99	<0.0001
N content (g) = a DBH^b	0.468	2.14	0.99	<0.0001

Data from Noguchi et al. (2009).

To elucidate the effects of thinning on aboveground biomass and C and N contents, a tree census was conducted in which the DBH of the *C. japonica* trees was measured. In March 2003, six 15 x 15 m study plots were established. A thinning practice (25% in trunk volume) was conducted in December 2003 in CS where three of the study plots were present. Another three plots without thinning were surveyed as control plots. At the beginning of the survey, the stand densities were 1850 and 2030 trees ha^{-1} in the thinned and control plots, respectively, reduced in the former to 890 trees ha^{-1} by thinning. In the tree census of this study, DBH of *C. japonica* trees in the plots were measured in March 2003, November 2003, December 2004 and December 2005, which were subjected to allometric equations to estimate the aboveground biomass and aboveground C and N contents.

The mean aboveground biomass prior to the thinning practice (November 2003) was 229 and 222 Mg ha^{-1} in the control plots and those for thinning, respectively. The aboveground biomass increased during the survey period and became 242 Mg ha^{-1} in the control plots in December 2005. In the thinned plots, conversely, thinning practice reduced the aboveground biomass by 30%, and the biomass was 174 Mg ha^{-1} in Dec. 2005 (2 years after thinning). The aboveground C and N contents showed similar trends in terms of temporal changes and thinning effects. In December 2005, the aboveground C contents were 115 and 83 Mg C ha^{-1} in the control and thinned plots, respectively, whereas the aboveground N contents were 535 and 371 kg N ha^{-1}, respectively (Table 11).

Table 11. Aboveground biomass and aboveground carbon and nitrogen storage in the thinned and the control plots. Data are Mean (SE) (N=3)

	Biomass		Carbon		Nitrogen	
	Mg ha^{-1}		Mg C ha^{-1}		Kg N ha^{-1}	
	Control	Thinned	Control	Thinned	Control	Tinned
Mar. 2003	221 (36)	214 (11)	105 (17)	102 (5)	495 (71)	478 (29)
Nov. 2003	229 (37)	222 (12)	109 (18)	106 (6)	511 (73)	494 (31)
Thinning in Dec. 2003						
Dec. 2004	235 (38)	167 (17)	112 (18)	80 (8)	521 (75)	358 (36)
Dec. 2005	242 (38)	174 (18)	115 (18)	83 (8)	535 (74)	371 (38)

Data from Noguchi et al. (2009).

In the control plots, the accumulation rates of aboveground biomass were 5.8 to 8.3 Mg ha^{-1} yr^{-1}, as opposed to 7.0 to 7.9 Mg ha^{-1} yr^{-1} in the thinned plots. In the latter, the accumulation rates of aboveground biomass following thinning (7.4 Mg ha^{-1} yr^{-1}) were 94% of those prior to thinning (7.9 Mg ha^{-1}yr^{-1}), although the thinning decreased the aboveground biomass by 30%. Conversely, the accumulation rates of aboveground C ranged from 2.8 to 4.0 Mg C ha^{-1} yr^{-1} and from 3.4 to 3.8 Mg C ha^{-1} yr^{-1} in the control and thinned plots, respectively; those of N ranged from 10 to 16 and 13 to 16 kg N ha^{-1} yr^{-1}, respectively. The accumulation rates of aboveground C and N after thinning were approximately 90% of those prior to thinning in the thinned plots, although the C and N contents decreased by 30% as well as biomass. Thus, the accumulation rates of aboveground C and N were similar between the control and thinned plots during both the periods prior to and following the thinning practice.

To estimate the litterfall production rates, samples were collected monthly from March 2003 to June 2006 using litter traps. Each litter trap covered 0.5 m^2 and 3 litter traps were positioned within each 15 x 15 m plot. In this study, litterfall only from *C. japonica* (needle leaves, branches and reproductive organs) were used for analyses. To estimate the C and N deposition through litterfall, samples collected from April 2003 to March 2004 were pooled and analyzed for C and N concentrations using an NC analyzer (Sumigraph N22F).

The annual litterfall production rates in the thinned plots (2.2 to 2.8 Mg ha^{-1} yr^{-1}) were approximately 55% of those in the control plots (3.9 to 5.2 Mg ha^{-1} yr^{-1}), although the aboveground biomass of *C. japonica* in the thinned plots was 70% of those in the control plots following the thinning practice. The annual deposition rates of C and N through litterfall in the thinned plots were also approx. 55% of those in the control plots. The largest fraction of the litterfall was needle leaves, which accounted for more than 80% of dry matter, C and N of litterfall (data not shown). A difference in litterfall production rates between the plots was also observed in 2003, the data for which were partly obtained prior to the thinning (July to November 2003). This is probably because the amounts of litterfall produced prior to thinning accounted for less than 10% of the annual production (data not shown).

In this study, trees felled by the thinning practice were not removed and were left on the forest floor, meaning they also contributed to C and N depositions in this *C. japonica* plantation, in which the amounts of dry matter, C and N supplied to the forest floor were 63 Mg DW ha^{-1}, 30 Mg C ha^{-1} and 151 kg N ha^{-1}, respectively.

In this study, the trees felled by the thinning practice were mainly small-sized. As a result, the thinning decreased aboveground C and N contents in the thinned plots by 30%, where the number of trees was 50% of those prior to thinning (Table 11). Conversely, the accumulation rates of aboveground C and N contents were similar in the control and thinned plots (Table 12). This is probably because the growth rates of the felled trees were slow due to limited light resources under the canopy of taller trees and their limited contribution to stand growth rates. Another possible reason is that the growth of residual trees might be improved following the thinning (Sheriff, 1996).

The effects of thinning on the deposition rates of C and N through litterfall appeared to be more evident than the contents and accumulation rates of aboveground C and N, in which the deposition rates in the thinned plots were 55% of those in the control plots (Table 13). This would result from reduced leaf biomass at the stand level by thinning and might also result from the retention of leaves in the lower canopy of residual trees due to improved light resource availability by thinning (Medhurst and Beadle, 2001). Felled trees are also a source of C and N deposition to the forest floor. In this study, the amounts of C (30 MgC ha^{-1}) and N (151 kgN ha^{-1}) of the felled trees were 10 and 2.5 times larger than the C and N contained in the litterfall, respectively (Tables 11 and 13). Thus, the felling of trees by thinning practices would be a major source of C and N deposition to the forest floor. However, the effects of the felled trees on C and N dynamics should be carefully examined, because the decomposition rates are slower for stems, a major component of felled trees, than for leaves, a major component of litterfall.

Table 12. Increment of aboveground biomass and aboveground carbon and nitrogen storage in the thinned and the control plots. Data are Mean (SE) (N=3)

Year	Biomass		Carbon		Nitrogen	
	Mg $ha^{-1}yr^{-1}$		Mg C $ha^{-1}yr^{-1}$		Kg N $ha^{-1}yr^{-1}$	
	Control	Thinned	Control	Thinned	Control	Thinned
2003	8.3 (0.9)	7.9 (0.9)	4.0 (0.5)	3.8 (0.4)	16 (2)	16 (2)
Thinning in Dec. 2003						
2004	5.8 (1.4)	7.7 (1.3)	2.8 (0.7)	3.7 (0.6)	10 (2)	15 (3)
2005	7.5 (0.1)	7.0 (1.2)	3.6 (0.1)	3.4 (0.6)	13 (1)	13 (2)

Data from Noguchi et al. (2009).

Table 13. Production rates of litterfall and deposition rates of carbon abd nitrogen through litterfall. Data are Mean (SE) (N=3)

Year	Litterfall		Carbon		Nitrogen	
	Mg ha^{-1} yr^{-1}		Mg C ha^{-1} yr^{-1}		Kg N ha^{-1} yr^{-1}	
	Control	Thinned	Control	Thinned	Control	Thinned
2003	5.0 (0.4)	2.8 (0.5)	2.5 (0.2)	1.4 (0.2)	43 (4)	24 (4)
2004	5.2 (1.2)	2.7 (0.5)	2.6 (0.6)	1.3 (0.3)	45 (11)	24 (4)
2005	3.9 (0.6)	2.2 (0.2)	1.9 (0.3)	1.1 (0.1)	33 (6)	20 (2)

Period of one year were July to June of next year. Data modified from Noguchi et al. (2009).

Conclusion

We studied the seasonal changes of material transformation and the effects of thinning on the transformations in *C. japonica* stand in a watershed in eastern Japan, and obtained the following results:

1. Nitrogen fixation during litter decomposition was exceptionally high in *C. japonica* litter, and the N increase in the *C. japonica* litter affected the subsequent N dynamics and decomposition patterns.
2. The fluxes of CO_2, CH_4 and N_2O differed in terms of the slope position, affected by the soil temperature and soil water content.
3. The annual rates of lead and cadmium deposition in the KHC still have been higher than those in remote areas. In this forest ecosystem,

accumulation of lead and cadmium in the A_0 layer was limited because of the rate of litter decomposition and the effective internal cycling through vegetation. Lead isotope analyses revealed that atmospherically-derived lead was distributed mostly in the surface soil (0–10 cm).

4. Nitrogen in green needles returned by thinning has a function whereby the N is temporarily stored and released later because the N immobilization and fixation dominated compared to the mass loss of the needles during the early stage decomposition process.
5. The accumulation rates of aboveground C and N were apparently unchanged following the thinning practice although the contents and deposition rates of aboveground C and N were decreased by thinning.

ACKNOWLEDGMENTS

We appreciate the Ibaraki District Forest Office of Kanto Regional Forest Office for giving us permission to use the national forest for this study. We would also like to thank the staff of the Forest Technology Center, Kanto Regional Forest Office. This research was financially supported by Grants-in-Aid for Scientific Research (#18580156, # 18780049) from the Japanese Society for the Promotion of Science (JSPS), and by a Research Grant (#200003) from FFPRI, Japan.

REFERENCES

Aastrup, M., Iverfeldt, Å., Bringmark, L., Kvarnäs, H., Thunholm, B., and Hultberg, H. (1995). Monitoring of heavy metals in protected forest catchments in Sweden. *Water Air Soil & Pollution*, 85, 755–760. Adriano, D. C. (2001). *Trace Elements in Terrestrial Environments*. New York, NY:Springer.

Aiba, Y., Haibara, K., and Kawabata, S. (1983). The effects of intensive tending work on soil productivity (II) Decomposition and movement of potential nutrients of fallen sugi (*Cryptomeria japonica* D. DON) foliage from green pruning. *Journal of Japanese Forestry Society,* 65, 215–219 (in Japanese with English summary).

Asami, T. (2001). *Toxic metals pollution in Japanese soils.* Tokyo:Agune-gijutsucenter (in Japanese).

Avila, A. and Rodrigo, A. (2004). Trace metal fluxes in bulk deposition, throughfall and stemflow at two evergreen oak stands in NE Spain subject to different exposure to the industrial environment. *Atmospheric Environment.* 38, 171–180.

Baker, T. G. and Attiwill, P. M. (1984). Acetylene reduction in soil and litter from pine and eucalypt forests in south-eastern Australia. *Soil Biology and Biochemistry*, 16, 241–245.

Berg, T., Røyset, O., and Steinnesm, E. (1994). Trace elements in atmospheric precipitation a Norwegian background stations (1989-1990) measured by ICP-MS. *Atmospheric Environment*, 28, 3519–3536.

Berg B. and McClaugherty, C. (2004). Plant litter decomposition humus. Springer-verlag, New York.

Berthelsen, B. O., Sterinnes, E., Solberg, W., and Jingsen, L. (1995), Heavy metal concentrations in plants in relation to atmospheric heavy metal deposition. *Journal of Environmental Quality,* 24, 1018–1026.

Bindler, R., Brännvall, M. L., Renberg, I., Emteryd, O., and Grip, H. (1999). Natural lead concentrations in pristine boreal forest soils and past pollution trends: A reference for critical load models. *Environmental Science and Technology,* 33, 3362–3367.

Borg, H. and Johansson, K. (1989). Metal fluxes to Swedish forest lakes. *Water, Air, and Soil Pollution,* 47, 427–440.

Campbbell, J. L., Hornbeck, J. W., Mitchell, M. J., Adams, M. B., Castro, M. S., Driscoll, C. T., Kahl, J. S., Kochnderfer, J. N., Likens, G. E., Lynch, J. A., Murdoch, P. S., Nelson, S. J., and Shanley, J. B. (2004). Input-output budgets of inorganic nitrogen for 24 forest watersheds in the northeastern United States: A review. *Water, Air, and Soil Pollution,* 151, 373–396.

Croteau, M.-N., Luoma, S. N., and Stewart, A. R. (2005). Trophic transfer of metals along freshwater food webs: Evidence of cadmium biomagnification in nature. *Limnology and Oceanography,* 50, 1511–1519.

DesJardins, M. J., Teler, K., and Beauchamp, S. (2004). Apportioning atmospheric pollution to Canadian and American sources in Kejimkujik National Park, Nova Scotia, using Pb isotopes in presuoutation. *Atmospheric Environment,* 38, 6875–6881.

Duke, L. D., Lo, T. S., and Turner, M. W. (1999) Chemical constituents in storm flow vs. dry weather discharges in California storm-water conveyances. *Journal of the American Water Resources Association / AWRA,* 35, 821–836.

Durand, P., Neal, C., Jeffery, H. A., Ryland, G. P., and Neal, M. (1994). Major, minor and trace element budgets in the Plynlimon afforested catchments (Wales): general trends, and effects of felling and climate variations. *Journal of Hydrology*, 157, 139–156.

Duvigneaud, P. and Denaeyer-de Smet, S. (1970). Biological cycling of minerals in temperate deciduous forests. In: D. E. Reichle (Ed.), *Analysis of Temperate Forest Ecosystems* (pp. 199–255). New York, NY: Springer-Verlag.

Forest Soils Division (1972). Classification of Forest soil in Japan. *Bulletin of Government Forest Experiment Station,* 280, 1–28 (In: Japanese with English summary).

Frei, M., Bielert, U., and Heinrichs, H. (2000). Effects of pH, alkalinity and bedrock chemistry on metal concentrations of springs in an acidified catchment (Ecker Dam, Harz Mountains, FRG). *Chemical Geology,* 170, 221–242.

Friedland, A. J. and Johnson, A. H. (1985). Lead distribution and fluxes in a high-elevation forest in northern Vermont. *Journal of Environmental Quality,* 14, 332–336.

Friedland, A. J., Craig, B. W., Miller, E. K., Herrick, G. T., Siccama, T. G., and Johnson, A. H. (1992). Decreasing lead levels in the forest floor of the Northeastern USA. *AMBIO,* 21, 400–403.

Galloway, J. N., Thornton, J. D., Norton, S. A., Volchok, H. L., and Mclean, R. A. (1982). Trace metals in atmospheric deposition: A review and assessment. *Atmospheric Environment,* 16, 1677–1700.

Galloway, J. N., Aber, J. D., Erisman, J. W., Seitzinger, S. P., Howarth, R. W., Cowling, E. B., and Cosby, J. B. (2003). The nitrogen cascade. *Bioscience,* 53, 341–356.

Gélinas, Y. and Schmit, J. P. (1998). Estimation of the bulk atmospheric deposition of major and trace elements to a rural watershed. *Atmospheric Environment,* 32, 1473–1483.

Haack, U. K., Heinrichs, H., Gustsche, F. H., and Plessow, K. (2003). The isotopic composition of anthropogenic Pb in soil profiles of northern Germany: evidence for pollutant Pb from a continent-wide mixing system. *Water, Air, and Soil Pollution,* 150, 113–134.

Halstead, M. J. R., Cunninghame, R. G., and Hunter, K. A. (2000). Wet deposition of trace metals to a remote site in Fiordland, New Zealand. *Atmospheric Environment*, 34, 665–676.

Hansmann, W. and Köppel, V. (2000). Lead-isotopes as tracers of pollutants in soils. *Chemical Geology*. 171: 123–144.

Harada, H., Satoo, H., Hotta, I., Hatiya, K., and Tadaki, Y. (1972). *Bulletin of Government Forest Experiment Station,*249*,* 17–74 (in Japanese with English summary).

Hashimoto, S., Morishita, T., Sakata, T., Ishizuka, S., Kaneko, S., and Takahashi, M. (2011). Simple model for soil CO_2, CH_4, and N_2O fluxes calibrated using a Bayesian approach and multi-site data. *Ecological modelling,* 220*,* 1283–1292.

Heinrichs, H. and Mayer, R. (1977). Distribution and cycling of major and trace elements in two central European forest ecosystems. *Journal of Environmental Quality,* 6*,* 402–407.

Hernandez, L., Probst, A., Probst, J. L., and Ulrich, E. (2003). Heavy metal distribution in some French forest soils: evidence for atmospheric contamination. *The Science of the Total Environment,* 312, 195–219.

Hibbs, D. E. and Cromack, K. Jr. (1990). Actinorhizal plants in Pacific Northwest forests. In: C. R. Schwintzer, and J. D. Tjepkema (Eds.), *The biology of* Frankia *and actinorhizal plants* (pp. 343–363). San Diego, CA: Academic Press.

Hirai, K., Noguchi, K., Mizoguchi,T., Kaneko, S., and Takahashi, M. (2007). Contribution of subsoil and season for nitrogen mineralization under field condition in forest soil. *Japanese Journal of Forest Environment,* 49*,* 51−59 (in Japanese with English summary). Hirai, K., Noguchi, K., Yamanaka, T., Kaneko, S., and Takahashi, M. (2011). Nitrogen supply from decomposing green needle after non-commercial thinning at a sugi (*Cryptomeria japonica*) stand. *Japanese Journal of Forest Environment,* 53*,* 17−22 (in Japanese with English summary).

Hornbeck, L. W., Bailey, S. W., Buso, D. C., and Shanley, J. B. (1997). Streamwater chemistry and nutrient budgets for forested watersheds in New England: variability and management implications. *Forest ecology and management,* 93, 73–89.

Inagaki, Y., Yoshinaga, S., Yamada, T., Shinomiya, Y., and Torii, A. (2009). Mass loss and nitrogen release of wood debris remaining after thinning in Japanese cedar and hinoki cypress plantations. *Japanese Journal of Forest Environment* 51*,* 63−67 (in Japanese with English summary).

IPCC Climate change 2007 (2007). *The scientific basis.* Cambridge: Cambridge University Press.

Ishizuka, S., Sakata, T., Sawata, S., Ikeda, S., Takenaka, C., Tamai, N., Sakai, H., Shimizu, T., Kan-na, K., Onodera, S., Tanaka, N., and Takahashi, M.

(2006). High potential for increase in CO_2 flux from forest soil surface due to global warming in cooler areas of Japan. *Annals of Forest Science,* 63, 537–546.

Itoh, Y., Takahashi, M., Akama, A., Mizoguchi, T., Yoshinaga, S., Tsurita, T., and Abe, T. (2005). Lead and cadmium cycling in a small forest catchment in Japan'. *Phyton,* 45, 437–442.

Itoh,Y. and Yoshinaga, S, (2006). Atmospheric lead and cadmium deposition within forests in the Kanto district, Japan. *Journal of Forest Research,* 11, 137−142.

Itoh, Y., Noguchi, K., Takahashi, M., Okamoto, T., and Yoshinaga, S.(2007) Estimation of lead sources in a Japanese cedar ecosystem using stable isotope analysis. *Applied Geochemistry,* 22, 1223–1228.

Jeffries, D. S. and Snyder, W. R. (1981). Atmospheric deposition of heavy metals in central Ontario. *Water, Air, and Soil Pollution*, 15, 127–152.

Johnson, C. E., Siccama, T. G., Driscoll, C. T., Likens, G. E., and Moeller, R. E. (1995). Changes in lead biogeochemistry in response to decreasing atmospheric inputs. *Ecological Applications, 5,* 813–822.

Kabata-Pendias, A. (2001). Soil-plant transfer of trace elements — an environmental issue. *Geoderma,* 122, 143–149.

Kaste, J. M., Friedland, A., Probst, J. L., and Sturup, S. (2003). Using stable and radioactive isotopes to trace atmospherically deposited Pb in montane forest soils. *Environmental Science and Technology,* 37, 3560–3567.

Keith, H., Mackey, B. G., and Lindenmayer, D. B. (2009). Re-evaluation of forest biomass carbon stocks and lessons from the world's most carbon dense forests. *Proceedings of the National Academy of Sciences of USA,* 106, 11635–11640.

Kobayashi, J. (2001). Distribution of metals in precipitation in Kanazawa city in a 15-month period. *Journal of Health Science,* 47, 502–507.

Lawlor, A. J. and Tipping, E. (2003). Metals in bulk deposition and surface waters at two upland locations in northern England. *Environmental Pollution*, 121, 153–167.

Likens, G. E., Bormann, F. H., Pierce, R. S., Eaton, J. S., and Johnson, N. M. (1977). *Biogeochemistry of a Forested Ecosystem.* New York, NY: Springer-Verlag.

Likens, G. E. and Bormann, E. H. (1995). *Biogeochemistry of a forested ecosystems* (2nd ed). New York, NY:Springer-Verlag.

Lindberg, S. E. and Harriss, R. C. (1981). The role of atmospheric deposition in an eastern U.S. deciduous forest. *Water, Air, and Soil Pollution*, 16, 13–31.

Lindberg, S. E. and Turner, R. R. (1988). Factors influencing atmospheric deposition, stream export, and landscape accumulation of trace metals in four forested watersheds. *Water, Air, and Soil Pollution,* 39, 123–156.

Medhurst, J. L. and Beadle, C. L. (2001). Crown structure and leaf area index development in thinned and unthinned *Eucalyptus nitens* plantations. *Tree Physiology,* 21, 989–999.

Miller, E. K. and Friedland, A. J. (1994). Lead migration in forest soils: response to changing atmospheric inputs. *Environmental Science and Technology,* 28, 662–669.

Morishita, T., Aizawa, S., Yoshinaga, S., and Kaneko, S. (2011). Seasonal change in N_2O flux from forest soils in a forest catchment in Japan. *Journal of Forest Research,* 16, 386–393.

Morishita, T., Sakata, T., Takahashi, M., Ishizuka, S., Mizoguchi, T., Inagaki, Y., Terazawa, K., Sawata, S., Igarashi, M., Yasuda, H., Koyama, Y., Suzuki, H., Toyota, N., Muro, M., Kinjo, M., Yamamoto, H., Ashiya, D., Kanazawa, Y., Hashimoto, T., and Umata, H. (2007). Methane uptake and nitrous oxide emission in Japanese forest soils and their relations to soil and vegetation types. *Soil Science and Plant Nutrition,* 53, 678–691.

Mukai, H., Furuta, N., Fujii, T., Ambe, Y., Sakamoto, K., and Hashimoto, Y. (1993). Characterization of sources of lead in the urban air of Asia using ratios of stable isotopes. *Environmental Science and Technology, 27,* 1347–1356.

Mulder, E. G. and Brotonegoro, S. (1974). Free-living heterotrophic nitrogen fixing bacteria. In: A. Quispel (Ed.), *The biology of nitrogen fixation* (pp. 37–85). Amsterdam: North-Holland.

Nioh, I. (1980). Nitrogen fixation associated with the leaf litter of Japanese cedar (*Cryptomeria japonica*) of various decomposition stages. *Soil Science and Plant Nutrition*, 26, 117–126.

Nioh, I. (1989). Changes of nitrogen-fixing activity and nitrogen-fixing bacteria in decomposing Japanese cedar (*Cryptomeria japonica*) leaf litter. *Japanese Journal of Soil Science and Plant Nutrition,* 60, 236–239 (In: Japanese).

Nioh, I. and Haruta, Y. (1986). Estimation of Amount of Asymbiotically Fixed Nitrogen in the Leaf Litter of Japanese Cedar (*Cryptmeria japonica*) *Journal of the Japanese Forest Society,* 88, 314-319.

Noguchi, K., Hirai, K., Takahashi, M., Aizawa, S.,Itoh, Y., Shigenaga, H., Nagakura, J., Inagaki, Y., Kaneko, S., Tsurita, T., and Yoshinaga, S. (2009). Effects of thinning on aboveground carbon and nitrogen dynamics at a sugi (Cryptomeria japonica) plantation in northern Kanto region, eastern Japan. *Bulletin of the Forestry and Forest Products Research Institute,* 8, 205 – 214 (In: Japanese with English summary).

Nowack, B., Obrecht, J. M., Schluep, M., Sculin, R., Hansmann, W., and Köppel, V. (2001). Elevated lead and zinc contents in remote alpine soils of the Swiss National Park. *Journal of Environmental Quality,* 30, 919–926.

Nriagu, J. O. (1990). Global metal pollution. *Environment,* 32, 7-30.

Nriagu, J. O. (1989). A global assessment of natural sources of atmospheric trace metals. *Nature,* 338, 47–49.

Nylund, J. E. and Wallander, H. (1992). Ergosterol analysis as a means of quantifying mycorrhizal biomass. In: J. R. Norris, D. J. Read and K. Varma (Eds.) *Methods in microbiology, vol.24: Techniques for the study of mycorrhiza* (pp. 77–88). London, Academic Press.

O'Connell, A. M., Grove, T. S., and Malajczuk, N. (1979). Nitrogen fixation in the litter layer of eucalypt forests. *Soil Biology and Biochemistry*, 11, 681–682.

Ono, K., Hirai, K., Morita, S., Ohse, K., and Hiradate, S. (2009). Organic carbon accumulation processes on a forest floor during an early humification stage in a temperate deciduous forest in Japan: Evaluations of chemical compositional changes by ^{13}C NMR and their decomposition rates from litterbag experiment. *Geoderma* 151, 351–356.

Ovington, J. D. and Madgwick, H. A. I. (1957). Afforestation and soil reaction. *Journal of Soil Science*, 8, 141–149.

Pacyna, J. M. and Pacyna, E. G. (2001). An assessment of global and regional emissions of trace metals to the atmosphere form anthropogenic sources worldwide. *Environmental Review,* 9, 269–298.

Perry, D. A. (1994). Forest ecosystem. Baltimore, MA: The Johns Hopkins University Press.

Raich, J. W. and Nadelhoffer, K. J. (1989). Carbon allocation in forest ecosystems: Global trends. *Ecology,* 70, 1346–1354.

Reimann, C., Koller, F., Frengstad, B., Kashulina, G., Niskavaara, H., and Englmaier, P. (2001). Comparison of the element composition in several plant species and their substrate from a 1,500,000-km2 area in Northern Europe. *The Science of the Total Environment,* 278, 87–112.

Sakata, T., Ishizuka, S., and Takahashi, M. (2004). A method for measuring fluxes of greenhouse gases from forest soils. *Bulletin of the Forestry and Forest Products Research Institute*, 3, 259–265.

Salamanca, E. F., Kaneko, N., Katagiri, S., and Nagayama, Y. (1998). Nutrient dynamics and lignocellulose degradation in decomposing Quercus serrata leaf litter. *Ecological Research*, 13, 199–210.

Sheriff, D. W. (1996). Responses of carbon gain and growth of Pinus radiata stands to thinning and fertilizing. *Tree Physiology,* 16, 527–536.

Sherrell, R. M., Boyle, E. A., Falkner, K. K., and Harris, N. R. (2000). Temporal variability of Cd, Pb isotope deposition in central Greenland snow. *Geochemistry Geophysics Geosystems,* 1: 1999GC000007.

Shima, K. and Chiba, K. (2002). Changes in biological nitrogen fixation activity and organic composition of leaf litter with decomposition of evergreen species. *Journal of Japanese Forestry Society*, 84, 37–40 (In: Japanese with English summary).

Shotyk, W., Cheburkin, A. K., Appleby, P. G., Fankhauser, A., and Kramers, J. D. (1997). Lead in Three Peat Bog Profiles, Jura Mountains, Switzerland: Enrichment Factors, Isotopic Composition, and Chronology of Atmospheric Deposition. *Water, Air, & Soil Pollution*, 100, 297–310.

Sirahata, H., Elias, R. W., and Patterson, C. C. (1980). Chronological variations in concentrations and isotopic compositions of anthropogenic lead in sediments of a remote subalpine pond. *Geochimica et Cosmochimica Acta,* 44, 149–162.

Smith, W. H. and Siccama, T. G. (1981). The Hubbard Brook ecosystem study: biogeochemistry of lead in a northern hardwood forest. *Journal of Environmental Quality,* 10, 323–333.

Sollins, P., Cline, S. P., Verhoeven, T., Sachs, D., and Spycher, G. (1987). Patterns of log decay in old-growth Douglas-fir forests. *Canadian Journal of Forest Research*, 17, 1585–1595.

Staff, H. and Berg, B. (1977). Mobilization of plant nutrients in a Scots pine forest mor in central Sweden. *Silva Fennica*, 11, 210–217.

Stahr, K., Zöttl, H. W., and Hädrich, F. R. (1980). Transport of trace elements in ecosystems of the Bärhalde watershed in the southern black forest. *Soil Science,* 130, 217–224.

Takahashi, M. (1996). Water soluble elements in decomposing Japanese cedar needle litter. *Soil Science and Plant Nutrition*, 42, 395–399.

Takeda, K., Marumoto, K., Minamikawa, T., Sakugawa, H., and Fujiwara, K. (2000). Three-year determination of trace metals and the lead isotope ratio in rain and snow depositions collected in Higashi-Hiroshima, Japan. *Atmospheric Environment,* 34, 4525–4535.

Tsumura, A. and Yamasaki, S. (1998). Background levels of trace elements in rain and river waters in Japan. *Radioisotopes,* 47, 46–55 (in Japanese).

Turner, R. S., Johnson, A. H., and Wang, D. (1985). Biogeochemistry of lead in McDonalds Branch Watershed, New Jersey Pine Barrens. *Journal of Environmental Quality,* 14, 305–314.

Ukonmaanaho, L., Starr, M., Mannio, J., and Airola, T. R. (2001). Heavy metal budgets for two headwater forested catchments in background areas of Finland. *Environmental pollution,* 114, 63–75.

Van Hook, R. I., Harris, W. F., and Henderson, G. S. (1977). Cadmium, lead, and zinc distributions and cycling in a mixed deciduous forest. *AMBIO,* 6, 281–286.

Watmough, S. A. and Dillon, P. J. (2003a). Base cation and nitrogen budgets for seven forests catchments in central Ontario, 1983-1999. *Forest Ecology and Management,* 177, 155–177.

Watmough, S. A. and Hutchinson, T. C. (2003b). Uptake of 207Pb and 111Cd through bark of mature sugar maple, white ash and white pine: a field experiment. *Environmental pollution,* 121, 39–48.

Watmough, S. A. and Hutchinson, T. C. (2004). The quantification and distribution of pollution Pb at a woodland in rural south central Ontario, Canada. *Environmental pollution,* 128, 419–428.

Weete, J. D. (1974). Fungal lipid biochemistry—distribution and metabolism. New York, NY: Plenum.

West, A. W., Grant, W. D., and Sparling, G. P. (1987). Use of ergosterol, diaminopimelic acid and glucosamine contents of soils to monitor changes in microbial populations. *Soil Biology and Biochemistry*, 19, 607–612.

Whittaker, R. H., Likens, G. E., Bormann, F. H., Eaton, J. S., and Siccama, T. G. (1979). The Hubbard brook ecosystem study: forest nutrient cycling and element behavior. *Ecology,* 60, 203–220.

Yamanaka, T., Hirai, K., Aizawa, S., Yoshinaga, S., and Takahashi, M. (2011). Nitrogen-fixing activity in decomposing litter of three tree species at a watershed in eastern Japan. *Journal of Forest Research*, 16, 1–7.

Yamashita, T., Tobita, H., and Takeda, H. (1998). Nitrogen dynamics of decomposing Japanese cedar and Japanese cypress litter in plantation forests. In: Sassa, K. (Ed.) *Environmental forest science* (pp. 215–221). Dordrecht: Kluwer.

In: Watersheds
Editors: G. L. Kuhn and J. R. Emery
ISBN 978-1-62081-057-6

Chapter 2

NATURAL AND HUMAN INFLUENCES ON THE INORGANIC WATER CHEMISTRY IN AN ARID LAND OF NORTHERN XINJIANG, CHINA

Bingqi Zhu[1,*], Jingjie Yu[a], Xiaoguang Qin[2], Patrick Rioual[3], Heigang Xiong[3] and Ziting Liu[2]

[1]Key Laboratory of Water Cycle and Related Land Surface Processes, Institute of Geographic Sciences and Natural Resources Research, Chinese Academy of Sciences, Beijing, China

[2]Key Laboratory of Cenozoic Geology and Environment, Institute of Geology and Geophysics, Chinese Academy of Sciences, Beijing, China

[3]Key Laboratory of Oasis Ecology, Ministry of Education, Urumqi, China

ABSTRACT

The influence of rock mineralogy, climate, relief and human activity on natural water composition and export as a result of weathering is a major scientific concern both at the local scale, because of the link between water chemistry and biodiversity, and at the global scale, in relation to global climate changes. This study evaluates the inorganic water qualities of the northern Xinjiang rivers in order to assess the

[*] Corresponding author. Bingqi Zhu. E-mail address: zhubingqi@igsnrr.ac.cn; bingqizh@mail.iggcas.ac.cn.

natural and anthropogenic factors that have influenced the inorganic water chemistry of natural water in arid environment. Wide spatial variations are geographically characterized in the dissolved solids (TDS) and water chemistry. A typical carbonate alkaline with Ca^{2+}, HCO_3^- and SO_4^{2-} ions dominated the major inorganic water composition. The water chemistry agrees well with the "rock dominance" mechanism with a TDS value of 80-600 mg/L and a Na/(Na+Ca) ratio of 0.1-0.6. Stoichiometric analyses suggest that most of the ions derive from multiple sources. Carbonate weathering is the primary source, followed by silicate weathering and evaporite dissolution. The effects of local pollution have somewhat greater contribution on the oases and central areas of the Zhungarer watershed. Most parts of rivers show a trend toward an increase in the dissolved load from the upper to lower reaches, which is primarily attributed to evaporation effect of climate along the water course. Variations and distributional patterns of the inorganic water chemistry show evident correlation with regional lithological distribution, terrain and precipitation gradients, indicating a strong influence of local geology, topography and climate. Using rainwater as a baseline, contributions from atmospheric precipitation and rock weathering to the tributary chemistry are roughly evaluated to be in the range of 2-39% (average 13%) and 59-98% (average 86%), respectively.

Keywords: Inorganic water chemistry; arid environment; geological weathering; atmospheric input; climatic effect; northern Xinjiang China

1. INTRODUCTION

One can has intuitive appreciation that natural water contains a variety of dissolved and suspended constituents. Mountain streams appear pure, farm creeks often are muddy with sediments, and drainages in limestone-rich regions are fertile while those containing only granitic rocks usually are less so. The rivers of heavily populated areas, and even the rain, are polluted by human activities. Many factors influence the composition of natural water, causing variation from place to place. Rain is one source of chemical inputs to rivers, and a stream flowing through a region of relatively insoluble rocks can be chemically very similar to rain water in its composition. Most streams and rivers contain much more suspended and dissolved material than is found in the rain, however. Ultimately, all of the constituents of natural water originate from dissolution of the earth's rocks. The dissolving of rocks commonly is the major determinant of river water chemistry locally as well, but this varies with geology and with the magnitude of inputs via other pathways including precipitation, volcanic activity, and pollution. Materials are concentrated by evaporation and altered by chemical and biological interactions within the stream. Unlike seawater, which is quite constant everywhere and can be approximated with an artificial standard, river water varies considerably in its chemical composition (Livingstone, 1963).

The influence of rock mineralogy, climate, relief and human activity on natural water composition and export as a result of weathering is a major scientific concern both as the local scale, because of the link between stream chemistry and biodiversity, and at the global scale, in relation to global climate changes (Nedeltcheva et al., 2006; Quevauviller and Umezawa, 2011). Climate change has impacts on water resources and subsequently on the sustainability of our environment. Global warming would accelerate the hydrologic cycle, increasing both precipitation and evapotranspiration rates (Waggoner, 1990; Miller et al., 1997). Dealing with the potential threat of global warming has increased the complexity of water resources planning in arid and semiarid regions. Freshwater resources are under severe stress in arid to semi-arid northern China. At present, much areas in northern China is in a similar situation as countries in the Middle East and northern and southern Africa (Miller, 2002): the freshwater resources are almost completely allocated and existing resources are suffering from a decreasing quality due to pollution caused by urbanization, industry, deforestation, mining, agriculture and energy use (e.g., Huizenga and Harmse, 2005). Considering the growth of the population and the required socio-economic development, it will be unlikely that water resources in northern China are sustainable under the present conditions.

Northern Xinjiang (NX) in northwestern China is the geographical centre of the Asian Continent. Most of its rivers originate from peripheral mountain glaciers and snow and drain areas of diverse geology and climatical conditions, with lower reaches in the desert environment (Figure 1). Geochemical studies of these waters are of considerable importance for a better understanding of the influences of the geology and meteorology of their drainage areas, weathering of crustal rocks, evaporation under arid climates and the impact of human activities on the chemistry of naturally occurring water. Until now, little chemical data were available for the northern Xinjiang rivers, particularly for several international rivers such as the Yili, Ermin and Erlqis rivers, although the hydrogeochemistry of their neighboring watersheds, the Hexi-Corridor-Alashan watershed to the east and the Tarim watershed to the south, has been systematically investigated (Zhang et al., 1995; Yang and

Williams, 2003; Zhu and Yang, 2007; Si et al., 2009; Zhu and Yang, 2010; Zhu et al., 2011). The purpose of this study was to evaluate the inorganic water chemistry of the NX watersheds for the period 2008 in order to assess the natural and anthropogenic factors that influence the inorganic water chemistry of these watersheds in space.

2. REGIONAL BACKGROUND

Northern Xinjiang in northwestern China is located right in the middle of the great continent central Asian basin system, which hardly have any ocean drainage and cover a vast area between Mongolia in the east and Kazakhstan in the west. The narrow, 2,000 km long and 4,000 - 7,000 m high Tianshan Mountains divides Xinjiang into the Zhungarer Basin in the north and the Tarim Basin in the south. Northern Xinjiang ranges between 78° and 90°E (longitude) and 42° and 50°N (latitude) and is approximately 603,000 km^2 in area. Elevations increase from <500 m above sea level (asl) in the centre of the Zhungarer Basin to >3000 m asl in the south and north ranges (Figure 1). The topography is generally flat in the centre plain and is rather cragged in the peripheral mountainous areas. The wide piedmonts and pediment plains are marked by the Gobi desert, grassland for herding and oasis areas with intensive agricultural activities.

One of the prime characteristics of the central Asian deserts is their extreme continentality. Northern Xinjiang is considered to be the most central and remote land of the world's largest continent. It lies several thousands of miles away from any of the world's oceans. Strong oceanic influences like the monsoon are systematically blocked off by the Tibet Plateau and a number of very high ranges in the southwest. The Zhungarer Basin is, moreover, enclosed by the Tianshan Mountains in the south, the Altai in the northeast and the Alatau and Tarbagatei in the northwest. At a latitude of between 42-50°, the region clearly belongs to the temperate zone controlled by an arid and cold temperate continental climate, which could have prevailed in central Asia at least since the latest Oligocene (Sun et al., 2010). It is subjected to the dry arctic influence of the Mongolian anticyclone in winter, and receives most of its moisture from the Atlantic and the Mediterranean in summer. The climate throughout the inland region is, additionally, determined by the highly differentiated altitude of its terrain and the local air movement between the mountains and plains. The mean annual air temperature is about 5°C, with a minimum of -10°C to -20°C in January and a maximum of 28°C to 33°C in

August. The regional precipitation has a mean annual precipitation rate of 60-150 mm in the centre plain and 200-500 mm in the surrounding mountainous areas (Figure 1b). While the potential evapotranspiration is approximately 1000-3500 mm per year, it varies in aridity along the elevation gradient across northern Xinjiang, due to the distribution patterns of precipitation, temperature and relative humidity and the orientation of delivery of moisture by the westerlies.

The northern Xinjiang can be divided into three drainage systems according to the hydrological classification in China (Ma, 2002) (Figure 1a). Two of them, the Zhungarer and Yili watersheds, are inland continental drainage, and the third, the Erlqis watershed, drains into the Arctic Ocean. Much of the rivers in these watersheds display dendritic to sub-dendritic drainage patterns elongated in an E-W direction, except for the rivers originating from the north slope of the Tianshan Mountains, with an approximately S-N orientation (Figure 1a). Water resources in northern Xinjiang are mainly distributed in Yili (170.9×10^8 m^3), Arltai (129×10^8 m^3) and Tacheng districts (61.6×10^8 m^3) and the total volume of the surface water in northern Xinjiang is 435.7×10^8 m^3. The temporal distribution of runoff in a year is extremely heterogeneous with 50-70% in summer, 10-20% both in spring and autumn, and less than 10% in winter. The uneven spatial-temporal distribution of water resources and dry climate environment determine the complete dependence of oasis agricultural activities on water.

The Zhungarer Basin is a structural depression filled with unconsolidated Quaternary and Tertiary sediments, as much as 500-1000 m in thickness (XETCAS et al., 1978). The basin is also an extension of the Palaeozoic Kazakhstan block, which is surrounded by Palaeozoic folded mountains. Aeolian deposits are widespread in this basin. The Gulbantonggute Desert, the second largest sand sea in China (48,800 km^2), is located in the central part of the basin. The large geographic distance from the surrounding oceans and the presence of the rain-shadow effect due to the surrounding orographic conditions are responsible for the arid climate of the Zhungarer Basin, which is necessary for the development of the desert.

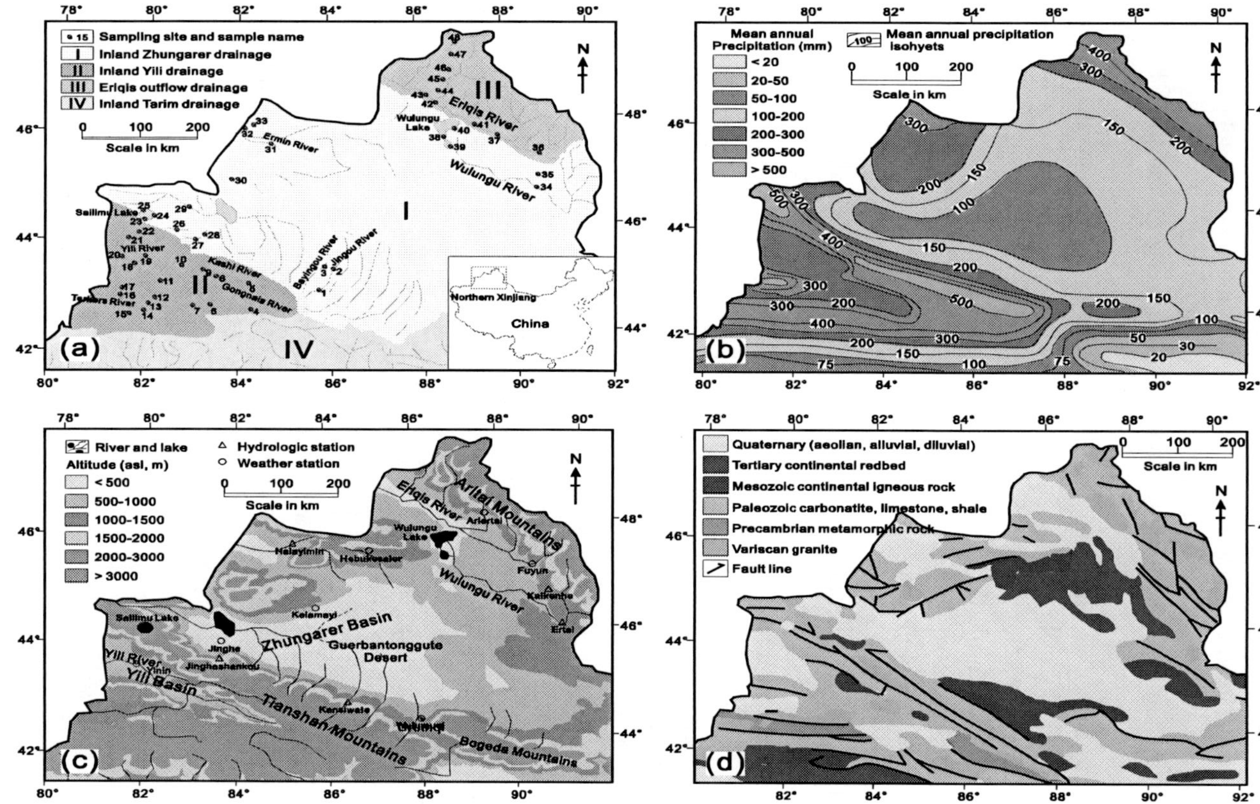

Figure 1. Hydrological setting and sampling locations map (a), mean-annual-precipitation-isohyets distribution map (b), topographical map (c) and lithological-distribution map (d) of northern Xinjiang.

Rivers in the Yili watershed converged into the Yili River at their lower reaches. The Yili River is a large international river running from its sources in the northeast Borohoro Mountains and southeast Halik Mountains, both branches of the eastern Tianshan Mountains, and flowing eastward through the Yili basin into Kazakstan (Figure 1). It drains igneous and metamorphic Precambrian (An€) and Variscan (γ_4) granite, Carboniferous (C_1, C_{2+3}) carbonatite and limestone and Quaternary sediments (Ma, 2002) (Figure 1d). The Yili River is about 430 km long inside the Yili watershed and contains three major tributaries in the catchment: the Kashi River in the northeast, the Gongnais River in the east and the Terkers River in the south and southeast. Much of the drainage areas of the tributaries are dominated by the sedimentary, mostly carbonate, lithologies (Figure 1d).

The Erlqis watershed has headwaters in the southern slopes of the Arltai Mountains (Figure 1a). Rivers converge into the Erlqis River at their lower reaches, and the Erlqis River is also a large international river, flowing through Kazakstan northward through Russia and finally into the Arctic Ocean. The stem channel of the Erlqis River is about 500 km long inside the Erlqis watershed. The Buerjin River, one of the most important tributaries of the Erlqis River, is about 250 km long and originates in the glaciers of the Youyi Mountains (4373 m asl). The lithologic outcrops in the catchment are mainly the Devonian marine carbonate, clastite (D2, D2+3), and the igneous Variscan (γ_4) granite and Quaternary sediments (Ma, 2002) (Figure 1d).

3. Methodology

The study water samples were collected under natural flow conditions in August 2008. In this month, runoff volumes of rivers in northern Xinjiang are almost under a yearly high-flow conditions due to high precipitation and air temperature, indicating a high influence of mountain glaciers and snow and climate on river hydrology. So it can be a suitable period to explore the geological and climatic effects on water hydrochemistry. Samples were collected in one-litre polyethylene bottles, which had been rinsed with deionized water and then rinsed thrice with raw water before sampling. Water samples were taken from various parts of the Yili, Zhungarer, and Erlqis watersheds (Figure 1a), including river stems, stream channels, hill slopes, wells, lakes, ponds, man-made trenches and reservoirs, as described in detail in Table 1. The sampling sites were chosen carefully in order to maximise representation of the diverse ecohydrological environments within the river systems. Taking into account that the tributaries reflect a much broader variety of geological, biological, and population patterns than do main stem rivers (Pawellek et al., 2002), it was of interest to sample tributary water to look for common features reflected in their hydrochemistry. Some pristine streams draining forested catchments were chosen in the Tianshan and the Arltai Mountains. All waters were surface-sampled except for well water extracted from pump shafts. Most samples were colourless, but a few were yellow, gray or turbid in colour due to dissolved iron content or suspended solid particles. The coloured river water samples primarily drained through the south and north piedmont areas of the Tianshan Mountains, with deep coves of loess and loess-like soil.

Table 1. Statistical summary of the parameters determined in the studied water samples (TH, total hardness, SAR, sodium adsorption ratio)

Parameter	Maximum	Minimum	Average	Median	10th	90th	SD
pH	9.81	7.00	-	7.85	7.33	8.24	0.47
Eh	-21.0	-187	-72.6	-72.5	-94.6	-41.8	27.6
EC(μs/cm)	5380	46.3	864	352	162	3426	1258
TDS(mg/L)	6200	24.6	580	186	86.1	1843	1071
Li^+(mg/L)	0.13	nd	0.02	0.01	nd	0.04	0.02
Na^+(mg/L)	1673	1.66	105	6.86	2.60	272	281
NH_4^+(mg/L)	0.17	nd	0.02	nd	nd	0.03	0.04
K^+(mg/L)	77.7	0.49	8.48	1.63	0.82	25.9	17.4
Mg^{2+}(mg/L)	190	0.63	27.2	6.49	2.44	81.8	49.5
Ca^{2+}(mg/L)	91.9	3.05	29.1	26.4	14.0	48.0	16.7
F^-(mg/L)	3.04	0.02	0.31	0.18	0.06	0.54	0.51
Cl^-(mg/L)	648	1.83	74.5	6.91	2.97	223	160
NO_3^-(mg/L)	36.5	nd	4.96	2.88	0.72	8.89	6.62
SO_4^{2-}(mg/L)	2655	2.52	205	45.3	9.01	682	437
HCO_3^-(mg/L)	799	9.76	117	69.4	29.7	283	142
TH(mg/L)	824	10.2	184	102	47.9	436	209
SAR	28.2	0.058	1.97	0.278	0.094	4.56	4.85

The parameters measured in this analysis include temperature (T), pH, electrical conductivity (EC), electrical potential (Eh), total dissolved solid (TDS), alkalinity, major cations (Li^+, Na^+, NH_4^+, K^+, Mg^{2+}, Ca^{2+}) and major anions (F^-, Cl^-, NO_2^-, Br^-, NO_3^-, $H_2PO_4^-$, SO_4^{2-}). T, pH, EC, Eh and TDS were measured in situ using a portable Multi-Parameter Analyzer (Eijkelkamp

18.28). The bias of the analytical procedures is <2% for pH and EP and <1% for EC, TDS and T. Cations and anions were determined by ion chromatography (IC, Dionex 600) with deionized water (EC<2.1 μS/cm) as the dilute base. The alkalinity was measured with a Hach digital titrator using the Gran method (Wetzel and Likens, 2000). Statistical summaries of the analytical data of physical parameters (pH, EC, Eh and TDS) and major ion concentrations are listed in Table 1. Total cations and anions (in unit of meq/L) are related by the equation $TZ^+ = 1.0094 \times TZ^- \pm 0.0218$, with a correlation coefficient of 0.9994 and a SD of ±4%. Total ions (TI, in unit of mg/L) and TDS are related by the equation: $TDS = 1.0249 \times TI - 6.6935$, with a correlation coefficient of 0.9991 and a SD of ±5%.

4. Inorganic Water Chemistry of the Northern Xinjiang – General Characterization

The pH values range between 7.0 and 9.81 with an average of 7.87 (Table 1 and Figure 2a), being neutral to mildly alkaline in nature. The slightly alkaline pH of 8+ is associated with weathering of fresh silicate rocks producing relatively large amounts of bicarbonate (e.g., White et al., 1999).

The TDS ranges between 24.6 mg/L and 6200 mg/L (Figure 2a), varying over two orders of magnitude from fresh (TDS < 1,000 mg/L) to brackish (1,000 mg/L < TDS < 10,000 mg/L) in water salinity. The median TDS concentration is 580 mg/L, which is slightly lower than the median of 625 mg/L of natural waters in southern Xinjiang (Zhang et al., 1995; Zhu and Yang, 2007), more than two times higher than the Changjiang-River (the largest river in China) median of 206 mg/L (Chen et al., 2002), and eight times the global median of 65 mg/L (Meybeck and Helmer, 1989). In general, the Zhungarer watershed had the highest TDS values among three watersheds, and the Erlqis watershed had the lowest value (Figure 2a).

Calcium is the dominant cation in most of water samples (Figs. 2b and 3), followed by sodium, magnesium and potassium, in that order. Bicarbonate is the major anion (Figures 2b and 3), followed by sulphate and chloride. There is a general trend towards HCO_3-predominance in fresh waters and SO_4^{2-} predominance in saline waters. The variations of the major-ion concentrations in river waters from upstream to downstream show no systematic variation for most solutes, except for Cl^- and Na^+, which display a preferential distribution of the higher concentration downward.

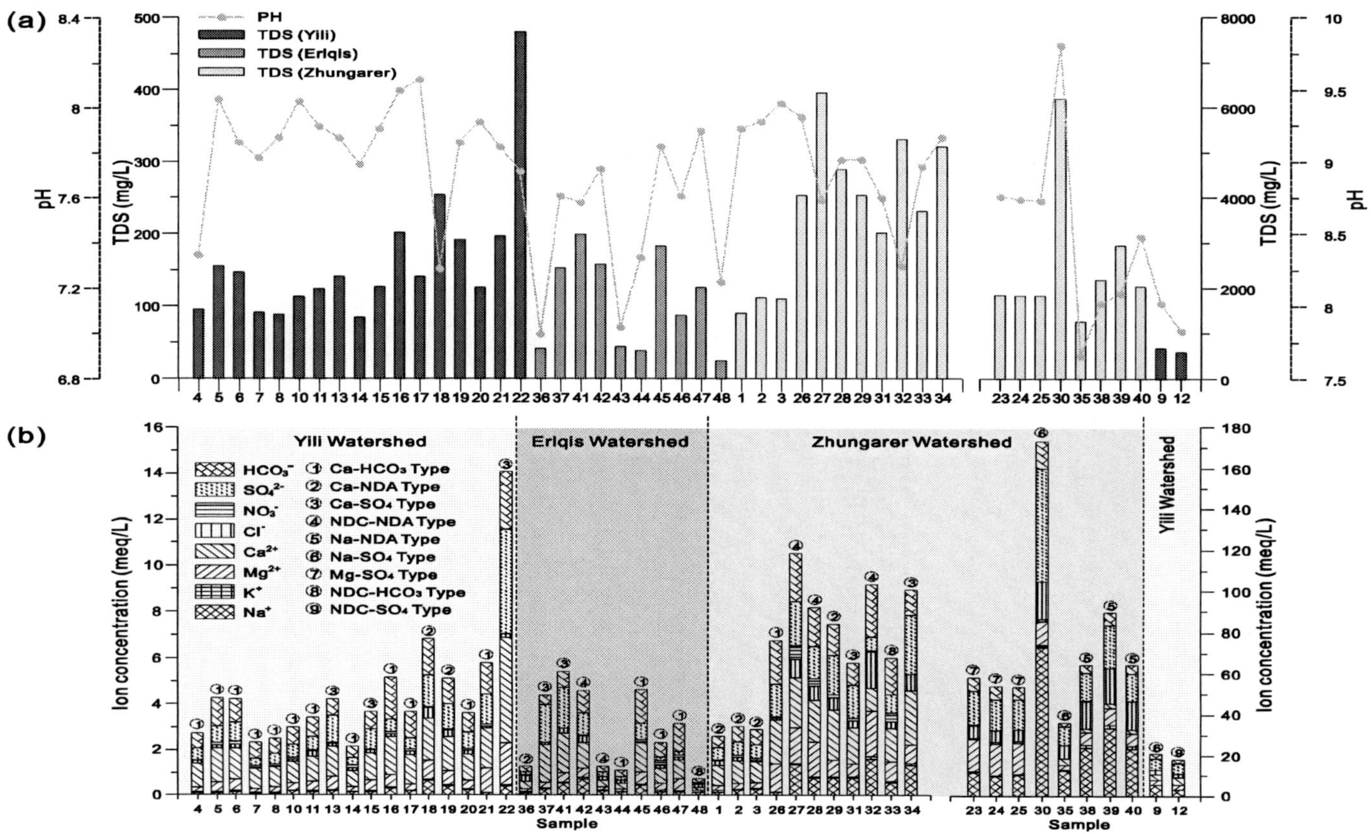

Figure 2. Histograms of TDS and pH (a) and ion concentrations (b) in natural waters in northern Xinjiang.

The chemical compositions of the natural waters in Piper's diagram (Figure 3) are plotted in fields where alkaline earths exceed alkalis and strong acids and weak acids dominate partially in composition, and these compositions can be classified into nine water types (Figure 2b) in Piper's terms (Piper, 1944). There are four main water types: Ca-HCO_3, Ca-NDA (non-dominant anion, Dissanayake and Weerasooriya, 1985), Ca-SO_4 and NDC-NDA (non-dominant cation, Dissanayake and Weerasooriya, 1985) or Na-NDA (Figures 2b and 3). Ca-type water is most widely distributed. The Ca-HCO_3 and Ca-NDA types mainly occur in the montane areas and the low-pitched piedmont zones. The Ca-SO_4 and Na-SO_4 types are mainly distributed in the transition band between oasis and desert plain. Mg-type water merely occurs in the Sailimu Lake.

The water type displays a slightly variational trend of zonal distribution, from the mountainous area to the pediment, oases and desert plain, in a rough corresponding order of Ca-HCO_3 (NDC-HCO_3), Ca-NDA, NDC-NDA, Ca-SO_4 and Na-SO_4 types. There is no Cl-dominated water type that occurs in the study area (Figure 3). Cl-type water is usually one of the main water types in the desert environment in northern China, such as in the Taklamakan Desert (Zhang et al., 1995; Zhu and Yang, 2007; Zhu et al., 2011) and the Badain Jaran Desert (Yang and Williams, 2003; Zhu and Yang, 2010; Zhu et al., 2011). Because no surface water exists in the hinterland of the Gulbantonggute Desert, no surface water sample was collected from the region in this work. Based on hydrochemical studies in the neighboring watersheds and the primarily survey of XETCAS et al. (1965) for groundwater in Xinjiang, one can infer that Cl-type water may be distributed in the Gulbantonggute Desert.

The sum of cation concentrations (TZ^+), a good indicator of weathering (Meybeck, 2005), varies from 0.36 meq/L to 45.7 meq/L (mean value 6.84 meq/L), except for Sample 30 (86.5 meq/L). Compared to the global rivers average (TZ^+=0.725 meq/L, Meybeck, 2005), rivers in northern Xinjiang are significantly enriched in dissolved solids, similar to those in southern Xinjiang (Zhang et al., 1995; Zhu and Yang, 2007), being typical for rivers in arid climatic conditions (e.g., Plant et al., 2001; Huizenga and Harmse, 2005). The dominated Ca-type in classification of natural waters in northern Xinjiang show a feature similar to the majority of large rivers on earth, which are typically carbonate rivers and alkaline in nature (Meybeck, 1987).

Compared with the upper ranges of typical values for major world rivers and the lower ranges for major pristine rivers (most common natural concentrations – MCNC) (Figure 4), such as the Amazon and Mackenzie (Meybeck and Helmer, 1989; Meybeck, 1996), the mean major ion concentrations of the studied waters are almost in excess of typical values of the major world rivers (Figure 4), except for Ca^{2+} and HCO_3^-. This suggests that the natural waters in northern Xinjiang fall close to the upper limit of concentrations in major world rivers, classified as “salted” by Meybeck (1996). Furthermore, compared with the chemical compositions of global rivers, the studied waters are notably characterised by an enormous range over three orders of magnitude for concentrations and yields (Figure 4). This indicates a definable spatial variation of hydrochemistry in northern Xinjiang watersheds.

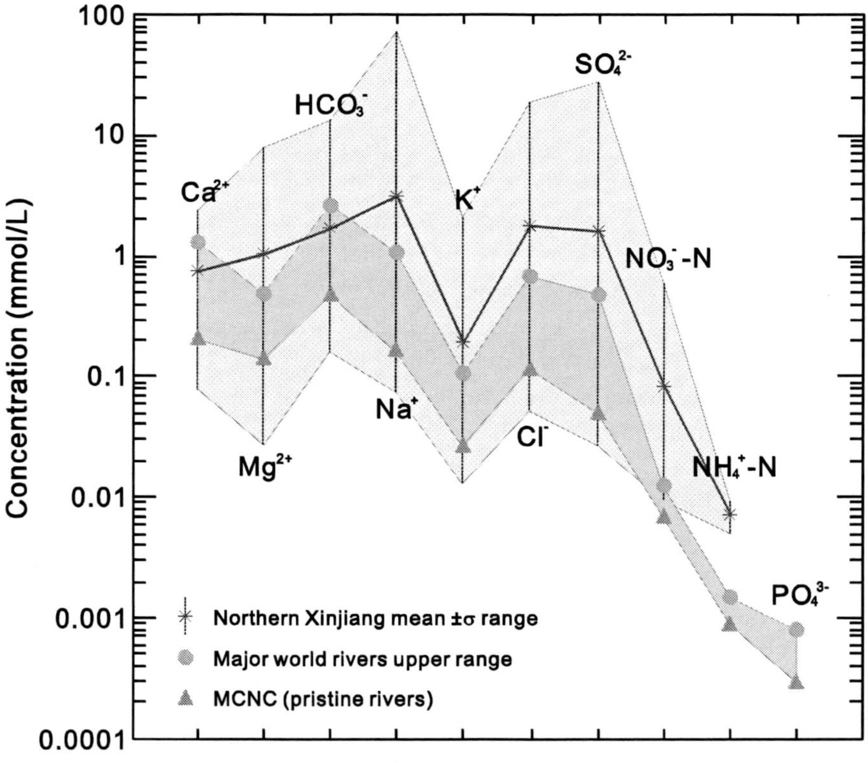

Figure 4. Typical ranges for major ions in the natural waters in three watersheds in northern Xinjiang. Also marked are the upper limits for these components in major world rivers and the median concentrations in pristine rivers (MCNC) (Meybeck and Helmer, 1989; Meybeck, 1996).

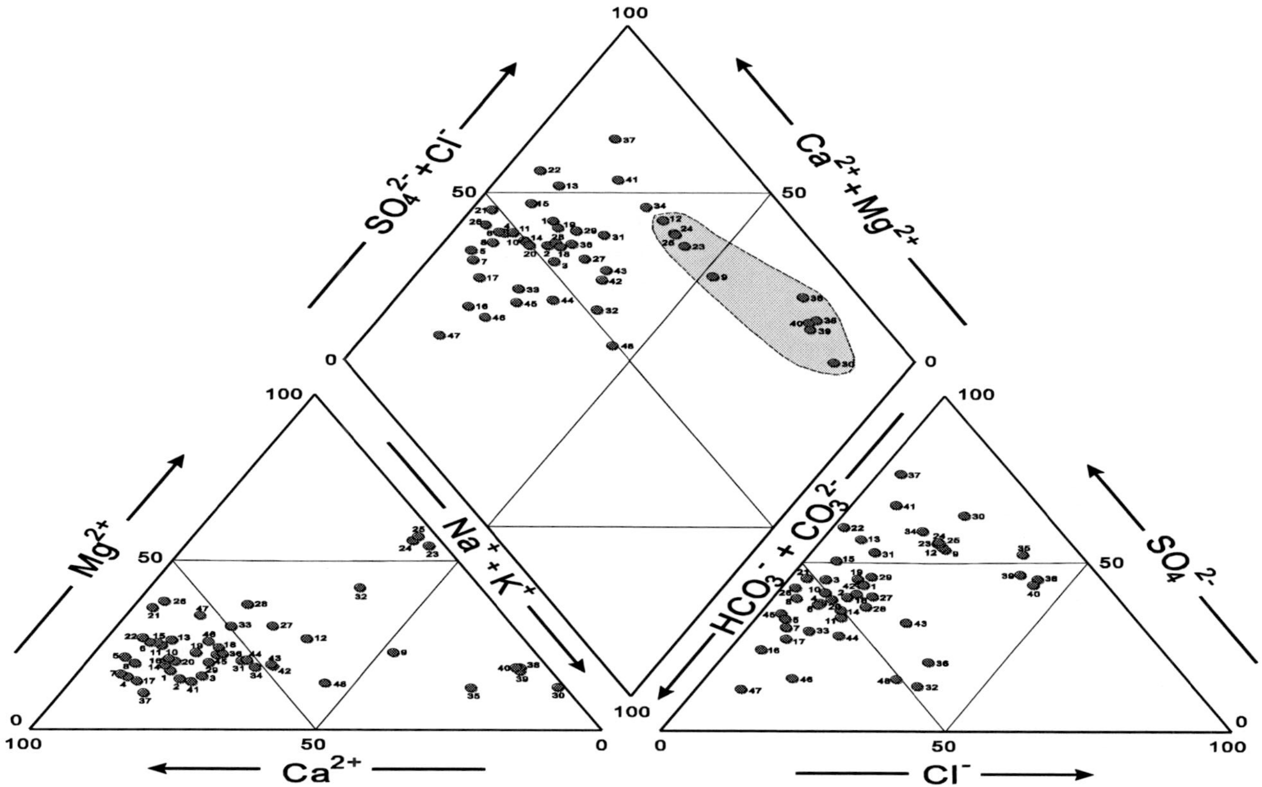

Figure 3. Piper plot of the study samples in three watersheds in northern Xinjiang. The shadow area represents the samples with highest TDS shown in the right side in Figure 2a,b.

5. Factors Controlling the Inorganic Water Chemistry

Natural water chemistry has been described in many geochemical studies as being determined essentially by a variety of physical, chemical and biological processes in the drainage basin, by which the major sources of soluble elements are introduced into natural water. These include primarily the following: (1) sea salt carried inland in the atmosphere and deposited in the river (cyclic salts), (2) the weathering of silicate, carbonate, evaporite and sulphide minerals, and (3) anthropogenic input (Berner and Berner, 1996). The types, elemental ratios and stoichiometry of solutes in water are thought to have origin implications related to the mechanisms described above (Gibbs, 1970; Feth, 1971; Stallard and Edmond, 1981, 1983; Kilham, 1990; Webster et al., 1994; Wu and Gibson, 1996; Meybeck, 2005).

5.1. Atmospheric Input

Recognizing the atmospheric source of dissolved materials in river water is essential for the identification of weathering sources within river systems (Stallard and Edmond, 1981). The average residence time of sea salts in the atmosphere is about 3 days (Junge, 1972). As a result, the contribution of cyclic salts to riverine-dissolved salt loads is expected to decrease with increasing distance from the source region of precipitation (Stallard and Edmond, 1981; Chen et al., 2002). Chloride is the most useful reference for marine input to river water (Stallard and Edmond, 1981; Meybeck, 1989), given that dry and wet depositions from a marine environment are important sources of halogens, which have less adsorbability to soils than metallic elements do (Tagami and Uchida, 2006). Apart from marine input, soil salt dissolution is another main source of chloride in river water. Chloride in surface waters in the absence of terrestrial sources declines systematically as a function of increasing distance from the source region (Stallard and Edmond, 1981).

The significant regions of precipitation in northern Xinjiang, due to the geographical and topographical conditions, are focused on the montane areas with high altitude (Figure 1b,c). Figure 5a illustrates that the annual mean summer precipitation (AMP) is strong positively correlated with the sampling site altitude. Chloride concentrations of the study waters, considered along with their altitude and AMP of sampling sites (Figure 5b,c), do not decline with increasing distance from the source region of precipitation, a clear difference from those rates observed for many world rivers (Stallard and Edmond, 1981). Instead, many of them show a contrary trend (Figure 5b,c), similar to the tributaries of the Changjiang watershed in China (Chen et al., 2002). These higher chloride concentrations in the downstream areas cannot be atmospheric (marine) in origin. Rather, they more likely originate from the dissolution of evaporites (halite) or are influenced by evaporation effect because chloride is a very conservative ion.

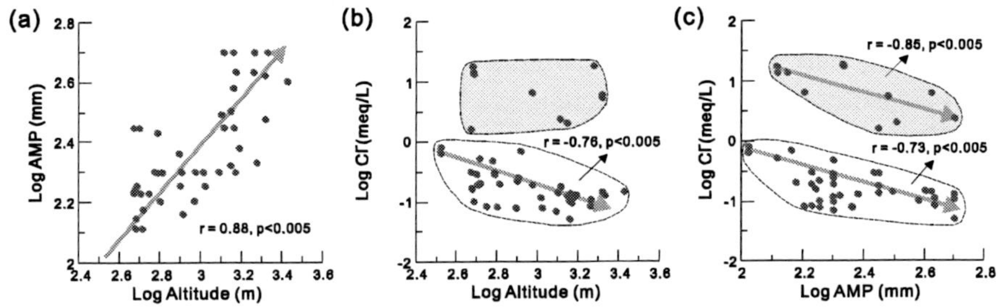

Figure 5. Annual mean precipitation (AMP) vs. altitude (a), and Cl^- concentration vs. altitude (b) and AMP (c) of the sampling sites in three watersheds in northern Xinjiang. The shadow circle represents the samples with highest TDS shown in the right side in Figure 2a, and the empty circle represents the samples with low TDS shown in the left side in Figure 2a.

5.2. Rock Weathering

Gibbs (1970) proposed that a simple plot of TDS versus the weight ratio of $Na^+/(Na^++Ca^{2+})$ or $Cl^-/(Cl^-+HCO_3^-)$ would provide information on the relative importance of three major mechanisms controlling natural water chemistry: atmospheric precipitation, rock weathering and evaporation and fractional crystallisation. Although Gibbs' method has proved to be controversial for waters with high $Na^+/(Na^++Ca^{2+})$ or $Cl^-/(Cl^-+HCO_3^-)$ (Feth, 1971; Stallard and Edmond, 1983; Kilham, 1990; Berner and Berner, 1996), it provides a simple tool to identify waters that are dominated by rock weathering (Berner and Berner, 1996). Plotting the corresponding data of northern Xinjiang in the Gibbs model revealed that the study waters are mainly characterised by a low ratio of $Na^+/(Na^++Ca^{2+})$ (0.1-0.6) (Figure 6a) and $Cl^-/(Cl^-+HCO_3^-)$ (0.05-0.5) (Figure 6b) and a moderate concentration of TDS (80-600 mg/L), typical of rock-weathering dominant waters (Gibbs, 1970). Taking rainfall as a baseline of regional hydrochemistry, Wu and Gibson (1996) suggested a Ca/Na versus TDS model to identify mechanisms controlling the water chemistry. Given the Waliguan precipitation as the representative of atmospheric “background” conditions of northern China (Tang et al., 2000; Zhao et al., 2008; Zhu and Yang, 2010), plotting the chemical data of the study waters in the Ca-Na-TDS diagram (Figure 6c) also revealed that rock weathering, particularly Ca- and Na-rock weathering, dominates the water chemistry in the majority of northern Xinjiang. The rock-weathering dominance of the major element chemistry provides a means of investigating further the weathering reactions in the three watersheds in northern Xinjiang because weathering of different parent rocks yields different combinations of dissolved cations and anions in solution (Meybeck, 1989; Meybeck, 2005).

Table 2. Correlation matrix of the dissolved ions in three watersheds in northern Xinjiang

	pH	TDS	Na^+	K^+	Mg^{2+}	Ca^{2+}	Cl^-	NO_3^-	SO_4^{2-}	HCO_3^-
pH	1.00									
TDS	0.59	1.00								
Na^+	0.44	0.95	1.00							
K^+	0.44	0.94	0.99	1.00						
Mg^{2+}	0.71	0.83	0.64	0.62	1.00					
Ca^{2+}	0.13	0.27	0.24	0.20	0.06	1.00				
Cl^-	0.44	0.95	1.00	0.99	0.63	0.23	1.00			
NO_3^-	-0.07	-0.13	-0.16	-0.18	-0.11	0.13	-0.16	1.00		
SO_4^{2-}	0.63	0.99	0.91	0.89	0.88	0.28	0.90	-0.16	1.00	
HCO_3^-	0.70	0.90	0.75	0.73	0.95	0.25	0.74	-0.04	0.93	1.00

A parametric correlation analysis for values of measured parameters is given in Table 2. HCO_3^- concentration is positively correlated with the concentrations of the base cations Mg^{2+} ($r=0.97$, $p<0.005$), Na^+ ($r=0.87$, $p<0.005$) and K^+ ($r=0.85$, $p<0.005$). Although the correlation coefficient between concentrations of HCO_3^- and Ca^{2+} for all the study samples is low ($r=0.49$, $p<0.005$), it is high between the two ions for most of the samples (Figure 7a). As a whole, the low correlation coefficient results from the disturbing effects of several abnormal samples, which had additional HCO_3^- or Ca^{2+} fractions that were responsible for other ions. For example, HCO_3^- concentration is better correlated with the concentrations of $Ca^{2+}+Mg^{2+}$ ($r=0.93$, $p<0.005$) (Figure 7b), and $Ca^{2+}+Mg^{2+}$ concentrations are better correlated with the concentrations of $HCO_3^{-}+SO_4^{2-}$ ($r=0.97$, $p<0.005$) (Figure 7c).

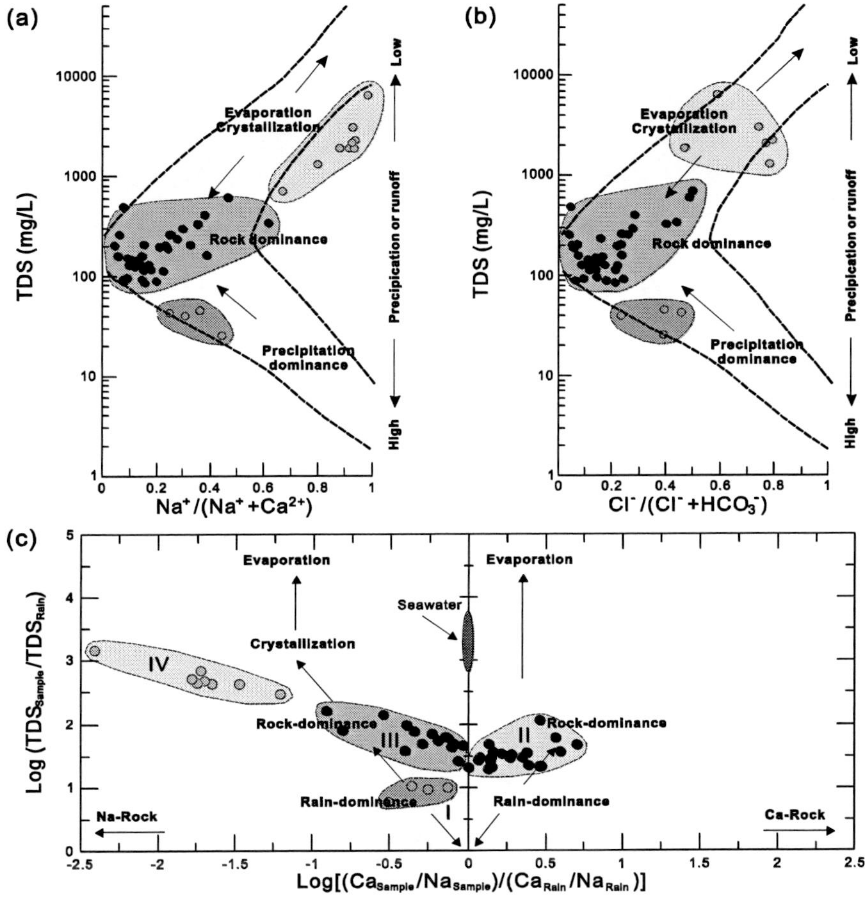

Figure 6. $Na^+/(Na^++Ca^{2+})$ (a) and $Cl^-/(Cl^-+HCO_3^-)$ weight ratios (b) vs. TDS content (after Gibbs, 1970), illustration of the Ca-Na-TDS model (c) (after Wu and Gibson, 1996) for natural waters in three watersheds in northern Xinjiang. The black circles in blue shadow represent the samples in rock dominance, the grey circles in yellow shadow represent the samples in evaporation-crystallization dominance and the empty circles in green shadow represent the samples in precipitation dominance. I, II, III and IV (in Figure 6c) represents the rain dominance, the Ca-rock dominance, the Na-rock dominance and the Crystallization dominance, respectively. The rainwater value was cited from the data of the Waliguan Station in northern China (Tang et al., 2000).

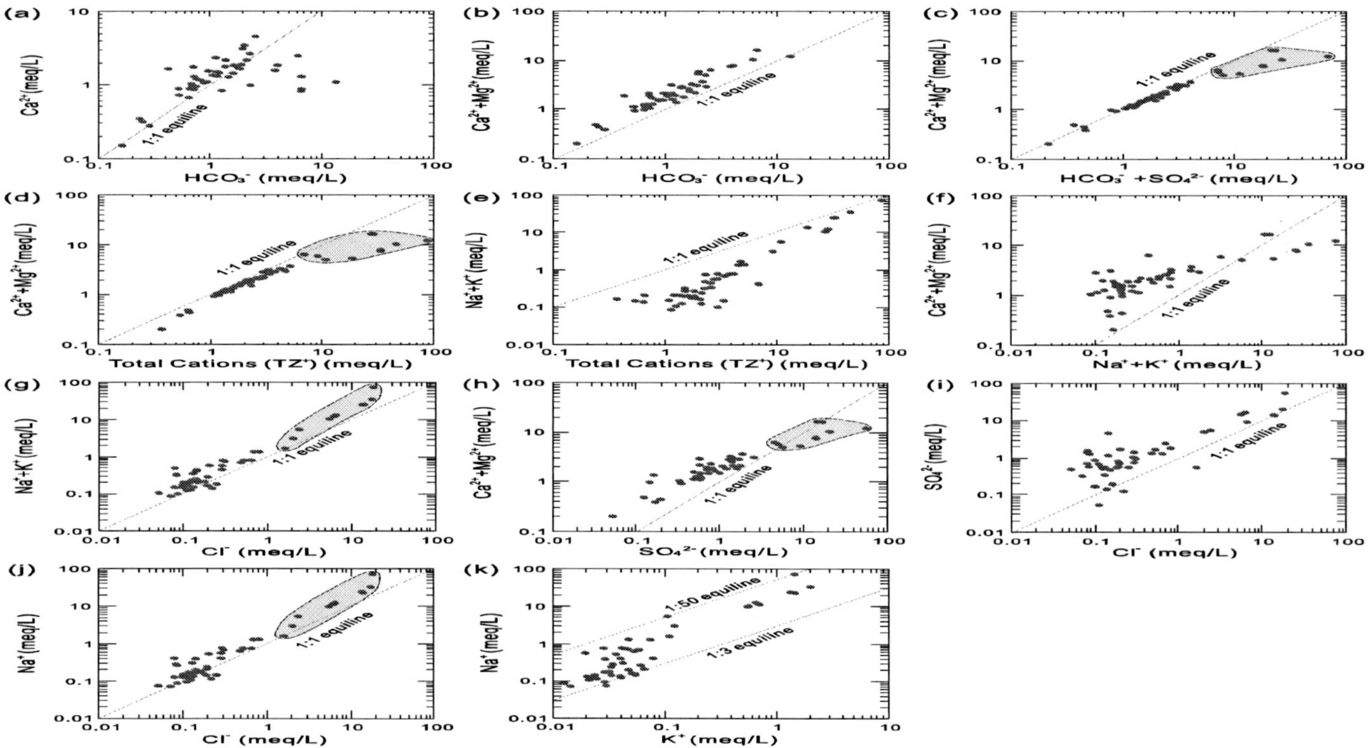

Figure 7. Scatter diagram of the concentrations of (a) Ca^{2+} vs. HCO_3^-, (b) (Ca^{2+}+Mg^{2+}) vs. HCO_3^-, (c) (Ca^{2+}+Mg^{2+}) vs. (HCO_3^-+SO_4^{2-}), (d) (Ca^{2+}+Mg^{2+}) vs. TZ^+, (e) (Na^++K^+) vs. TZ^+, (f) (Ca^{2+}+Mg^{2+}) vs. (Na^++K^+), (g) (Na^++K^+) vs. Cl^-, (h) (Ca^{2+}+Mg^{2+}) vs. SO_4^{2-}, (i) SO_4^{2-} vs. Cl^-, (j) Na^+ vs. Cl^-, (k) Na^+ vs. K^+ of the studied sample waters in three watersheds in northern Xinjiang. The shadow circle represents the samples with highest TDS values shown in the right side in Figure 2a.

The higher concentration of HCO_3^- in natural waters and its positive correlation with Ca^{2+} and Mg^{2+} concentrations can be attributed to their common origin in the weathering of carbonates (Stallard and Edmond, 1983; White, 2008). Subsequently, many river samples in this study maintain a stable molar Ca/Mg ratio of ~2, which is typical of most large rivers dominated by carbonate weathering (Meybeck, 1989; Pawellek et al., 2002). The plot of (Ca^{2+}+Mg^{2+}) vs. HCO_3^- content (Figure 7b) shows that the Ca^{2+}+Mg^{2+} contents of most samples are slightly in excess of HCO_3^- content (average ratio 1.2), suggesting that an excess of Ca^{2+}+Mg^{2+} in these waters should be balanced by other ions, such as SO_4^{2-} (Figure 7c) and Cl^- ($r=0.83$, $p<0.005$, between Ca^{2+}+Mg^{2+} and Cl^-). Most sample points approach the equal (1:1) trend with an average equivalent ratio of 0.73 in the plot of Ca^{2+}+Mg^{2+} vs. TZ^+ (Figure 7d). The low (Na^++K^+)/ TZ^+ ratio (about 0.26) (Figure 7e) and high (Ca^{2+}+Mg^{2+})/(Na^++K^+) ratio (about 5.63) (Figure 7f) reflect the dominance of carbonate weathering over silicate weathering as the major source of river solutes in northern Xinjiang. This is consistent with the lithological distributions of the study area (Figure 1d) that show that carbonate rocks are widespread in the three watersheds, particularly the Yili and Erlqis watershed (Figure 1d). Moreover, the more rapid dissolution of carbonate rocks compared to silicate (Meybeck, 1987; White, 2008) is likely what causes carbonate solute acquisition in an area with silicate rock distribution.

The comparison of chloride and sulphate contents to Na^++K^+ and Ca^{2+}+Mg^{2+} contents are often used as a reference for the evaporite source (Zhang et al., 1995; Chen et al., 2002). The Cl^- vs. Na^++K^+ comparison in Figure 7g displays an evident trend toward a 1:1 ratio, suggesting an important evaporite source (halite). The plot of Ca^{2+}+Mg^{2+} contents vs. SO_4^{2-} content (Figure 7h) also displays a ratio of Ca^{2+}+Mg^{2+} to SO_4^{2-} close to unity, indicating a common source from either carbonate weathering or the dissolution of gypsum for these solutes. Cl^-+SO_4^{2-} contents, in equivalent units on average, account for 55.7% (ranging from 16.6% to 88.6%) of the total anion concentration, with high concentrations occurring in the water from the low-altitude areas of the Zhungarer watershed. The strong correlation between Cl^- and SO_4^{2-} concentrations (Figure 7i, $r=0.95$, $p<0.005$) suggests a common source for these two elements. Case studies have also inferred that Cl^- and SO_4^{2-} seem to co-evolve in Chinese rivers (Hu et al., 1982), such as the Changjiang and Huanghe, possibly due to the formation of $CaSO_4$ (gypsum) being related to NaCl (halite). A few samples with (Ca^{2+}+Mg^{2+})/SO_4^{2-} < 1 in concentration ratios (Figure 7h) were presumably enriched in a Na_2SO_4 component (Glauber's salt).

The possible sources of Na^+ and K^+ concentrations in natural water are atmospheric precipitation, dissolution of rock salts (halite) and weathering of sodium-bearing minerals (Stallard and Edmond, 1983; Meybeck, 2005). The concentrations of Na^+ (0.07-72.8 meq/L, average 3.13 meq/L) and K^+ (0.01-1.99 meq/L, average 0.18 meq/L) vary largely throughout the three watersheds. Na, like Cl, is also a major component of sea water (cyclic salt). Considering the same variation trend between Na^+ and Cl^- (Figure 7j) in the study waters from upstream to downstream, Na^+ in water cannot originate from the contribution of cyclic salt. Further, the Na^+ concentration in the study waters is significantly in excess of the Cl^- concentration (Figure 7j), with the equivalent concentration ratios (>1) between Na^+/Cl^- generally higher than the sea-water value of 0.86, suggesting a source other than atmospheric precipitation for a great part of Na^+ in the study waters. If the halite dissolution process is responsible for the Na^+, the Na^+/Cl^- equivalent ratio should be approximately 1, whereas a Na^+/Cl^- ratio greater than 1 typically indicates that the Na^+ is released from silicate weathering (Meyback, 1987; Kumar et al., 2009). The average Na^+/Cl^- equivalent ration is 1.7:1 in the study waters, indicating a significant contribution from silicate weathering on Na^+. Na^+ and K^+ concentrations are highly correlated (Figure 7k, $r=0.95$, $p<0.005$, Table 2), but the concentration ratio of Na^+/K^+ declines with the decreasing concentrations from ~50:1 to ~3:1. This suggests that at low concentrations, a greater proportion of both Na^+ and particularly K^+ may originate from dissolution of silicate mineral, such as feldspar, illite or kaolinite (Pawellek et al., 2002).

5.3. Anthropogenic Input

Agricultural practices can pose a serious threat to groundwater quality, especially in arid regions because the growing season and its high water demands coincide with the dry period. The use of fertilisers with a strong polluting effect on groundwater is a widespread agricultural practice. The use of nitrogen has led to extremely high nitrate concentrations in the groundwater of many regions (Stigter et al., 1998). Compared with the nitrate contents (0.5-7.7 mg/L) in the study surface waters, several groundwater samples (Sample 27, 28, 33) collected from different irrigated oases in the Zhungarer watershed have evidently higher concentrations of nitrate (22-37 mg/L, Table 1). It indicates that the groundwater in these irrigated lands in the Zhungarer basin have been influenced greatly by the local agricultural fertilisation processes.

Three inorganic nitrogen species (nitrate, nitrite, ammonium) have different measured distributions in the study waters. Nitrite was not detected in any sample, and ammonium only occurred in a few samples in this study (Table 1). Nitrate was detected in most of the study samples, with clearly higher concentrations than the upper range for major world rivers (Figure 4). Its concentration also showed unsystematic variations with any ion concentrations and parameters (Table 2) or along river courses, suggesting that point sources, presumably of anthropogenic origin, play an important role. In comparison with the mean value (2.46 mg/L) of nitrate, somewhat higher concentrations of nitrate (7-37 mg/L) were observed in several samples collected from oases with dense population or farming, particularly groundwater and river water taken from agricultural and residential regions (Table 1). This indicates an evident anthropogenic input of nitrate in these places, probably from agricultural activities (fertilisers) or municipal and industrial discharges. These inputs are possibly modified by biological processes in the river itself, with oxidation of NH_4^+ to NO_2^- and subsequently to NO_3^- (Pawellek et al., 2002), which can explain the distribution patterns of three inorganic nitrogen species in the study waters. Concentrations of phosphorus species (orthophosphate, phosphate and hypophosphite), the limiting factors of autotrophic life, are not detected in any sample in this study.

5.4. Climate Effect

The principles and processes discussed so far are the direct and fundamental ones that determine the composition of natural water. The manner in which the processes operate, however, and their relative importance in any one environment are controlled by numerous nonchemical forces. The processes of rock weathering are strongly influenced by temperature and by amount and distribution of precipitation. The influence of climate on water quality goes beyond these direct effects. Climatic patterns tend to produce characteristic plant communities and soil types, and the composition of water draining from such areas could be thought of as a product of the ecologic balance.

5.4.1. Zonal Distribution Patterns of Hydrochemistry

Certain of the major ionic constituents of natural water are influenced more strongly than others by climatic effects. Bicarbonate, for example, tends to predominate in water in areas where vegetation grows profusely. Humid temperate climates and warm wet climates generally are the most favorable for growth of vegetation. Runoff from tropical rainforest areas commonly is low in dissolved-solids concentration. An arid climate is unfavorable for rapid rates of solvent erosion, but concentration of dissolved weathering products by evaporation can give rise to water that is high in dissolved-solids content. On the other hand, the occasional flood runoff in such regions can be very low in dissolved material because soluble weathering products are not available in major quantity.

For the regional distribution patterns of chemical water types in the study area, the relatively wet zones with AMP>200 mm in the three watersheds consist of the Ca-HCO_3, Ca-NDA and NDC-NDA types (Figures 1-3); and in the relatively dry zones with AMP<200 mm, however, the Ca-SO_4 and Na-SO_4 types predominate and the Na-Cl type occur in the central sandy desert. The Ca-HCO_3-dominated chemistry of river water in the relatively wet zones couples well with regional carbonate lithological distributions (Figures 1d, 3) and confirms that rock weathering and the regional lithologic distributions have a major effect on the evolution of river chemistry in the wet zones. The precipitation processes accelerate the rock weathering processes, as confirmed by White and Blum (1995). The water types in dry zones tend to be relatively uniform. Because the lithological distributions are not homogeneous, evaporation processes, which can result in the selective precipitation of solute fraction ($CaCO_3$) and subsequent changes in water chemistry as water moves downstream in the relatively arid zones (Gibbs, 1970; Kilham, 1990; Feller, 2005), could be responsible for the accumulation and evolution of sodium and sulphate salts. It indicates a major effect of climate rather than geology on the evolution of water chemistry in dry zones.

The highest TDS concentrations were observed at some lake depressions around the central Zhungarer watershed, possibly resulting from two factors: tectonic control, given the closed shape and limited outlet flow of the lakes contributing to salt accumulation in hydrological terms, and the high evaporation potential leading to salt concentration. With low salinities of TDS<50 mg/L (Figure 2), the purest waters were observed in streams in the Erlqis watershed and were closely related to melt waters of the recent thaw of mountain ice and snow because their temperatures *in situ* were measured at lower than 5°C during the summer sampling. In comparison with the others, they could be sufficiently dilute to discount any possibility of extensive mixing with exotic non-glacial inputs, as the chemical composition of melt waters are constantly changing both with time and with distance from the glacial source (Green and Canfield, 1984; Webster et al., 1994).

The strong relationship between precipitations and altitudes of the sampling sites (Figure 5a) can be described as AMP = 0.14 × altitude + 101.3 ($r=0.88$, $p<0.005$). The dissolved solids in most samples decreased in concentration both with the increasing altitude or AMP (Figures 5b,c and 8). The fact that rock types are not distributed according to the altitude (Figure 1c,d) implies that topography and temperature, in addition to rock type and weathering, affects water chemistry indirectly by influencing AMP/ evaporation processes.

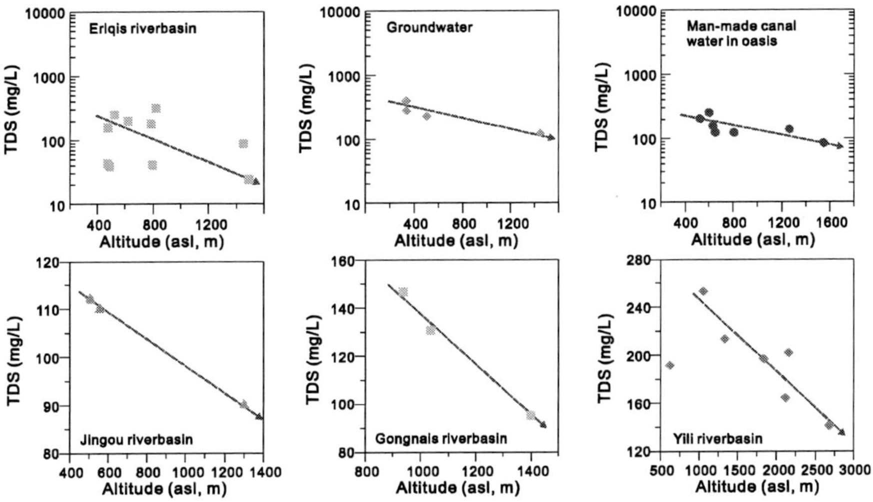

Figure 8. TDS vs. altitude (m, asl) of the sample waters from the three watersheds in northern Xinjiang.

5.4.2. Evaporation Effect

The studied water samples contain more dissolved solids than the major world rivers (Figure 4). The concentration of Cl^-, the lowest enriched major anion (mean value 76.5 mg/L) in the study waters (Figure 4), was higher than the mean value (8.1 mg/L) of world rivers (Tagami and Uchida, 2006). Given that the precipitation rate of approximately 100-500 mm/y (Figure 1b) in northern Xinjiang is less than the world average (880 mm/y, Tagami and Uchida, 2006), evaporation process must cause a climatic condensation effect on halogen concentrations. Additionally, the courses of northern Xinjiang rivers are usually longer than 300 km, they have low flow velocities (about 5 km/hr in average), and their lower reaches are often exposed to the desert environment (Figure 1). Thus, there is enough time for evaporation-crystallisation processes to facilitate condensation effects on the water chemistry of northern Xinjiang.

The dissolved solid loads in rivers such as the Erlqis, Wulungu, Jingou, Gongnais and Yili river increase from the upper to lower reaches (Figure 8). The river water at the middle reaches of the Jingou, Gongnais and Wulungu have TDS values of 90, 95 and 321 mg/L, respectively, and gradually change to higher values of 112, 147 and 2000 mg/L after flowing about 70, 100 and 300 km downward, respectively. Because the Jingou, Gongnais and Wulungu rivers have few tributaries over their water courses (Figure 1a), the water flowing along the lengths of these courses does not mix extensively with water from different sources. In addition, samples from a single river were collected on the same day, and therefore, the seasonal changes of discharge and sources cannot be the reasons for the variation in the dissolved solid concentrations. Accordingly, the increase of TDS from upper reaches toward lower reaches in these rivers can be interpreted in terms of evaporation and dissolution of soil salts along their river courses. If evaporation causes the concentration of dissolved material during water flow, given no change in the flowing speed of ~5 km/hr, it is intense enough in these drainage basins that runoff can decrease to one-third to one-fifth during one day.

5.5. Contributions from Precipitation and Weathering Input

Na^+ and Cl^- in water are both conservative ions. Cl^- in stream water in wet conditions is primarily influenced by climate (atmospheric precipitation) and not by geological weathering (Feller, 2005). Thus, the excess of Na^+ concentration in comparison to Cl^- concentration in stream water relative to the Na^+ concentration in local rain can be regarded as the contribution from regional weathering and anthropogenetic sources with rainfall as a baseline (Wu and Gibson, 1996).

A portion of the study waters, for which chemistries were illustrated above to be mainly controlled by rock-weathering dominance in Figure 6a-c, is mainly tributary stream water collected from montane areas with relatively high precipitation. The aridity and human activities in these places are relatively low, and certain vegetation is covered. Water chemistry will be less influenced by dissolution of evaporite, evaporation processes and anthropogenic source (Feller, 2005). According to this consideration, contributions from regional rock weathering and precipitation for various ions in the tributary water can be evaluated roughly from the following relations

(Dunne, 1978), respectively: (1) $\left[Cation\right]_{rain} = \frac{\left[Cl\right]_{stream} \times \left[Cation\right]_{sea}}{\left[Cl\right]_{sea}}$ and (2) $\left[Cation\right]_{weathering} = 1 - \frac{\left[Cl\right]_{stream} \times \left[Cation\right]_{sea}}{\left[Cl\right]_{sea}}$, where the brackets refer to average concentrations, and "rain" and "weathering" indicate the ion concentrations from atmospheric precipitation and rock weathering, respectively. Although some degree of uncertainty may exist in this approach, it effectively assesses the relative role of precipitation and weathering effects (Mast and Drever, 1990; Ahmad et al., 1998; Chae et al., 2004). Based on the above relations (1) and (2), the ionic contributions resulting from atmospheric precipitation to the tributary waters in the three watersheds are evaluated to be between 2% and 39% (average 13%) of the total dissolved solids, and between 59% to 98% (average 86%) from regional rock weathering. This indicates that rock weathering contributes to tributary chemistry more effectively than precipitation, but both are regionally variable in the study area.

Sample 47 (a spring-fed stream water) and Sample 48 (a glacier-fed lake water) were taken from the high region of the Karnas Geopark in the Erlqis watershed, only kilometres away from each other (Table 1, Figure 1a). The contribution of precipitation to Sample 48 (34%) is much higher than that to Sample 47 (7%), although Sample 47 has a clearly higher TDS value (126 mg/L) than Sample 48 (24.6 mg/L) (Table 1). To a certain extent, the mountain glacier water represents a "pure" atmospheric input and the spring water a "pure" geological weathering input due to its close rock-water interaction. The lower atmospheric contribution but higher TDS concentration of Sample 47, relative to that of Sample 48, also indicates the more significant effect of rock weathering than atmospheric precipitation on water chemistry.

CONCLUSION

The database in this study contains 13-ion analysis data sets of 48 sampling sites from three international watersheds (the Yili, Zhungarer and Erlqis) in northern Xinjiang. The majority of the study water is typical carbonate water and is alkaline in nature. Calcium is the dominant cation, and bicarbonate and sulphate are the dominant anions in most areas. Four major water types, Ca-HCO_3, Ca-NDA, Ca-SO_4 and NDC-NDA or Na-NDA type, were identified using the Piper model. Most of the sample waters agree well with the "rock dominance" field of the Gibbs model, with a TDS value of 80-

600 mg/L and a Na/(Na+Ca) ratio of 0.1-0.6, and indicate that the water chemistry in northern Xinjiang is greatly controlled by rock weathering. The high concentrations of alkali earth metals and alkalinity and the high (Ca+Mg)/(Na+K) and Na/Cl ratios indicate that the solute sources in water are largely derived from carbonate weathering and secondly from silicate weathering. The effects of local pollution are of minimal contribution to the montane and piedmont areas but of somewhat greater contribution to the oases and central areas of the Zhungarer watershed. Parts of some rivers show a trend toward increase in the dissolved load from upper to lower reaches, primarily attributed to evaporation or the dissolution of soil salts along the water course. The water type distributions and solute concentrations, varying in water between wet and dry regions and along changes in altitude, indicate a close relationship among regional lithological distribution, topography and hydrothermal conditions on the regional hydrochemical evolution. Using rainwater as a baseline, a conceptual evaluation shows the contributions from atmospheric precipitation and rock weathering to the amounts of dissolved solids in tributary waters approximately in the range of 2-39% (average 13%) and 59-98% (average 86%), respectively.

ACKNOWLEDGMENTS

This work was financially supported by the National Basic Research Program of China (973 Program, No. 2009CB421305), the National Natural Science Foundation of China (Grant no.: 40901059, 91025023, 40930105), and the Important Consultation Project of Chinese Engineering Academy "Strategic study of the water resource on the sustainable development of Xinjiang".

REFERENCES

Ahmad, T., Khanna, P.P., Chakrapani, G.J., Balakrishnan, S., 1998. Geochemical characteristics of water and sediment of the Indus River, Trans-Himalaya, India: constraints on weathering and erosion. *Journal of Asian Earth Sciences* 16(2-3), 333-346.

Berner, E.K., Berner, R.A., 1996. *Global Environmental: Water, Air and Geochemical Cycles*. Prentic-Hall, New Jersey.

Chae, G.T., Yun, S.T., Kim, K.H., Lee, P.K., Choi, B.Y., 2004. Atmospheric versus lithogenic contribution to the composition of first- and second-order stream waters in Seoul and its vicinity. *Environment International* 30, 73-85.

Chen, J., Wang, F., Xia, X., Zhang, L., 2002. Major element chemistry of the Changjiang (Yangtze River). *Chemical Geology* 187, 231-255.

Dissanayake, C.B., Weerasooriya, S.V.R., 1985. A geochemical classification of groundwater of Sri Lanka. *Journal of the National Science Council of Sri Lanka* 13(2), 147-186.

Dunne, T., 1978. Rates of chemical denudation of silicate rocks in tropical catchments. *Nature* 274, 244-246.

Feller, M.C., 2005. Forest harvesting and streamwater inorganic chemistry in western North America: a review. *Journal of the American Water Resources Association* 41(4), 785-811.

Feth, J.H., 1971. Mechanisms controlling world water chemistry: evaporation-crystallization process. *Science* 172, 870-871.

Gibbs, R.J., 1970. Mechanisms controlling world water chemistry. *Science* 170, 1088-1090.

Green, W.J., Canfield, D.E., 1984. Geochemistry of the Onyx River (Wright Valley, Antarctica) and its role in the chemical evolution of Lake Vanda. *Geochimica et Cosmochimica Acta* 48, 2457-2467.

Hu, M.H., Stallard, R.F., Edmond, J.M., 1982. Major ion chemistry of some large Chinese rivers. *Nature* 298, 550-553.

Huizenga, J.M., Harmse, J.T., 2005. Geological and anthropogenic influences on the inorganic water chemistry of the Jukskei River, Gauteng, South Africa. *South Africa Journal of Geology* 108, 439-447.

Junge, C., 1972. Our knowledge of the physico-chemistry of aerosols in the undisturbed marine environment. *Journal of Geophysical Research* 77, 5183-5200.

Kilham, P., 1990. Mechanisms controlling the chemical composition of lakes and rivers: data from Africa. Limnology and Oceanography 35(1), 80-83.

Kumar, S.K., Rammohan, V., Sahayam, J.D., Jeevanandam, M., 2009. Assessment of groundwater quality and hydrogeochemistry of Manimuktha River basin, Tamil Nadu, India. *Environmental Monitoring and Assessment* 159, 341-351.

Livingstone DA. 1963. Chemical composition of rivers and lakes. *Geological Survey Paper,* 440-G, pp 1-64.

Ma, L., 2002. *Geological Atlas of China.* Geological Press, Beijing. (in Chinese).

Mast, M.A., Drever, J.I., 1990. Chemical weathering in the Loch Vale watershed, Rocky Mountain National Park, Colorado. *Water Resource Research* 26, 2971-2978.

Meybeck, M., 1987. Global chemical weathering of surficial rocks estimated from river dissolved loads. *American Journal of Science* 287, 401-428.

Meybeck, M., 1996. River water quality: global ranges, time and space variabilities, proposals for some redefinitions. *Verhandlungen des Internationalen Verein Limnologie*, 26, 81-96.

Meybeck, M., 2005. Global occurrence of major elements in rivers. In: Drever, J.I. (Eds.), Treatise on Geochemistry: Surface and Ground Water, *Weathering, and Soils*. Elsevier, Amsterdam, pp. 207-223.

Meybeck, M., Helmer, R., 1989. The quality of rivers: from pristine stage to global pollution. *Palaeogeography, Palaeoclimatology, Palaeoecology,* 75, 283-309.

Miller, G.T., 2002. *Living in the Environment: Principles, Connections, and Solutions.* Wadsworth Publishing Company, Belmont, USA. 758pp.

Miller, K.A., Rhodes, S.L., Macdonnell, L.J., 1997. Water allocation in a changing climate: institutions and adaptation. *Climatic Change* 35, 157-177.

Nedeltcheva, T.H., Piedallu, C., Gegout, J.C., Stussi, J.M., Boudot, J.P., Angeli, N., Dambrine, E., 2006. Influence of granite mineralogy, rainfull, vegetation and relief on stream water chemistry (Vosges Mountains, north-eastern France). *Chemical Geology* 231, 1-15.

Pawellek, F., Frauenstein, F., Veizer, J., 2002. Hydrochemistry and isotope geochemistry of the upper Danube River. *Geochimica et Cosmochimica Acta* 66(21), 3839-3854.

Plant, J., Smith, D., Smith, B., Williams L., 2001. Environmental geochemistry at global scale. *Applied Geochemistry* 16, 1291-1308.

Piper, A.M., 1944. A graphic procedure in the geochemical interpretation of water-analyses. *Transactions of the American Geophysical Union* 25, 914-923.

Quevauviller, P., Umezawa, Y., 2011. Climate-change impacts on water chemistry. *Trends in Analytical Chemistry* 30(8), 1213.

Si, J., Feng, Q., Wen, X., Su, Y., Xi, H., Chang, Z., 2009. Major ion chemistry of groundwater in the extreme arid region northwest China. *Environmental Geology* 57, 1079-1087.

Stallard, R.F., Edmond, J.M., 1981. Geochemistry of the Amazon: 1. Precipitation chemistry and the marine contribution to the dissolved load at the time of peak discharge. *Journal of Geophysical Research* 86, 9844-9855.

Stallard, R.F., Edmond, J.M., 1983. Geochemistry of the Amazon: the influence of the geology and weathering environment on the dissolved load. *Journal of Geophysical Research 88*, 9671-9688.

Stigter, T.Y., Van Ooijen, S.P.J., Post, V.E.A., Appelo, C.A.J., Dill, A.M.M.C., 1998. A hydrogeological and hydrochemical explanation of the groundwater composition under irrigated land in a Mediterranean environment, Algarve, Portugal. *Journal of Hydrology* 208, 262-279.

Sun, J., Ye, J., Wu, W., Ni, X., Bi, S., Zhang, Z., Liu, W., Meng, J., 2010. Late Oligocene-Miocene mid-latitude aridification and wind patterns in the Asian interior. *Geology* 38(6), 515-518.

Tagami, K., Uchida, S., 2006. Concentrations of chlorine, bromine and iodine in Japanese rivers. *Chemosphere* 65, 2358-2365.

Tang, J., Xue, H., Yu, X., Cheng, H., Xu, X., Zhang, X., Ji, J., 2000. The preliminary study on chemical characteristics of precipitation at Mt. Waliguan. *Acta Scientiae Circumstantiae* 20(4), 420-425 (in Chinese with English abstract).

Waggoner, P.E., 1990. *Climate Change and U.S. Water Resources.* John Wiley & Sons, New York.

Webster, J.G., Brown, K.L., Vincent, W.F., 1994. Geochemical processes affecting meltwater chemistry and the formation of saline ponds in the Victoria Valley and Bull Pass region, Antarctica. *Hydrobiologia* 281, 171-186.

Wetzel, R.G., Likens, G.E., 2000. *Limnological Analyses,* 3rd ed. Springer, New York.

White, A.F., 2008. Quantitative approaches to characterizing natural chemical weathering rates. In: Brantley, S.L., Kubicki, J.D., White, A.F. (Eds.), *Kinetics of Water-rock Interaction.* Springer, New York, pp. 469-543.

White, A.F., Blum, A.E., 1995. Effect of climate on chemical weathering in watersheds. *Geochimica et Cosmochimica Acta* 59(9), 1729-1747.

White, A.F., Bullen, T.D., Vivit, D.V., Schulz, M.S., and Clow, D.W., 1999. The role of disseminated calcite in the chemical weathering of granitoid rocks. *Geochimica Cosmochimica Acta* 63, 1939-1953.

Wu, Y., Gibson, C.E., 1996. Mechanisms controlling the water chemistry of small lakes in northern Ireland. *Water Research* 30(1), 178-182.

XETCAS (Xinjiang Expedition Team of the Chinese Academy of Sciences), IGCAS (Institute of Geology of the Chinese Academy of Sciences),

XBCAS (Xinjiang Branch of the Chinese Academy of Sciences)., 1965. Groundwater in Xinjiang. Science Press, Beijing. (in Chinese).

XETCAS (Xinjiang Expedition Team of the Chinese Academy of Sciences), IGCAS (Institute of Geography of the Chinese Academy of Sciences), *DGBNU* (Department of Geography of Beijing Normal University)., 1978. Geomorphology in Xinjiang. Science Press, Beijing. (in Chinese).

Yang, X., Williams, M.A.J., 2003. The ion chemistry of lakes and late *Holocene desiccation in the Badain Jaran Desert*, Inner Mongolia, China. Catena 51, 45-60.

Zhang, J., Takahashi, K., Wushiki, H., Yabuki, S., Xiong, J., Masuda, A., 1995. Water geochemistry of the rivers around the Taklimakan Desert (NW China): crustal weathering and evaporation processes in arid land. *Chemical Geology* 119, 225-237.

Zhao, Z., Tian, L., Fischer, E., Li, Z., Jiao, K., 2008. Study of chemical composition of precipitation at an alpine site and a rural site in the Urumqi River Valley, Eastern Tien Shan, China. *Atmospheric Environment* 42, 8934-8942.

Zhu, B., Yang, X., 2007. The ion chemistry of surface and ground waters in the Taklimakan Desert of Tarim Basin, western China. *Chinese Science Bulletin* 52(15), 2123-2129.

Zhu, B., Yang, X., 2010. The origin and distribution of soluble salts in the sand seas of northern China. *Geomorphology* 123, 232-242.

Zhu, B., Yang, X., Rioual, P., Qin, X., Liu, Z., Xiong, H., Yu, J., 2011. Hydrogeochemistry of three watersheds (the Erlqis, Zhungarer and Yili) in northern Xinjiang, NW China. *Applied Geochemistry* 26, 1535-1548.

In: Watersheds
Editors: G. L. Kuhn and J. R. Emery

ISBN 978-1-62081-057-6

Chapter 3

SUMMER HOLDING AREAS OF THE GULF STURGEON WITHIN THE CONECUH/ESCAMBIA RIVER SYSTEM, ALABAMA AND FLORIDA

Paul M. Stewart[1,*], Jon B. Sawyer[2,†], Frank M. Parauka[3,‡] and Evelyn G. Reátegui-Zirena[4,§]

[1]Department of Biological and Environmental Sciences, Troy University, Troy, Alabama, US
[2]ARCADIS; Montgomery, Alabama, US
[3]U.S. Fish and Wildlife Service, Field Office; Panama City, Florida, US
[4]Department of Biological and Environmental Sciences, Troy University, Troy, Alabama, US

ABSTRACT

The Gulf of Mexico sturgeon *Acipenser oxyrinchus desotoi* is an anadromous subspecies of the Atlantic sturgeon *Acipenser oxyrinchus oxyrinchus*. This federally threatened fish once inhabited Gulf coastal rivers from the Mississippi River to Tampa Bay, but is now rarely found

[*] E-mail: mstewart@troy.edu.
[†] E-mail: bo.sawyer@arcadis-us.com.
[‡] E-mail: frank_parauka@fws.gov.
[§] E-mail: ereateguizirena@troy.edu.

south of the Suwannee River in Florida. The Conecuh/Escambia River system of Alabama and Florida contains more potential Gulf sturgeon spawning habitat than any other Florida Panhandle river, yet limited information is known about the sturgeons of this river. Radio transmitters were placed in 14 Gulf sturgeons collected from Escambia Bay during March through May of 2005. These radio-telemetered Gulf sturgeons were manually tracked via boat and radio receivers during April through November 2005 to collect summer holding area and movement data. Two, possibly three, summer holding areas were identified during this study, two in the Escambia River and one in the Delaney River.

INTRODUCTION

The Gulf of Mexico sturgeon *Acipenser oxyrinchus desotoi* (Vladykov, 1955) is a subspecies of the Atlantic sturgeon *A. o. oxyrinchus*. The Gulf sturgeon is an anadromous species that overwinters in the Gulf of Mexico and its associated estuaries, and then migrates into freshwater river systems of the Gulf Coast region for spawning during early spring (Huff, 1975). Gulf sturgeons remain within these rivers for summer refuge until they return to the feeding grounds of the Gulf of Mexico in autumn (Mason and Clugston, 1993).

The Gulf sturgeon historically inhabited coastal areas from the Mississippi River to Tampa Bay, but is now rarely found south of the Suwannee River in Florida (Foster and Clugston, 1997). Habitat destruction, overexploitation, and the blocking of spawning migration by dams and other anthropogenic river obstructions have all been major elements of the Gulf sturgeon's decline (Wooley and Crateau, 1985). Habitat destruction is primarily due to increased sediment loads within Gulf coastal plain rivers. Exploitation of the Gulf sturgeon increased in the 1860s, when it was discovered that smoked sturgeon meat could be used in place of smoked halibut (Smith, 1990). Another factor leading to increased harvesting of this fish was the discovery that its eggs could be used for high-quality caviar. To address the population decline of the Gulf sturgeon, it was federally listed as a threatened species in 1991; under the U.S. Endangered Species Act (USFWS, 1991). The U.S. Fish and Wildlife Service (USFWS) and the Gulf States Marine Fisheries Commission (GSMFC) formulated the Gulf Sturgeon Recovery and Management Plan in 1995 to identify priority action items needed to address numerous life history questions of this species (USFWS and GSMFC, 1995). Critical habitat was designated for the Gulf sturgeon in 2003 through a joint effort between the

USFWS and the National Marine Fisheries Service (USFWS and NMFS, 2003).

One very important element of the recovery plan established by the Gulf sturgeon Recovery/Management Task Team for the Southeast Region of the U.S. Fish and Wildlife Service (USFWS) is to locate and characterize critical habitat within each river that the Gulf sturgeon currently occupies (Hightower et al., 2002). Many studies recorded Gulf sturgeon congregations downstream of springs (Chapman and Carr, 1995; Foster and Clugston, 1997; Hightower et al., 2002). These areas may serve as thermal refuge from the surrounding higher temperatures within the rivers, and the deeper areas of river bottoms often associated with these springs may also serve as a refuge from surrounding river currents. Spawning habitats have been recorded within the Conecuh River, but only two areas have been suggested as summer holding areas within the Conecuh/Escambia River. One was suggested within the Conecuh River, near Brewton, Alabama, and the other was within the Escambia River, just below the Florida State Highway 184 Bridge (Quintette Bridge). Because the Conecuh/Escambia River has such large amounts of critical habitat, studying this river to obtain information about the location of additional summer holding areas and the Gulf sturgeon's use of habitat is an important part of implementing a successful recovery/management plan. This project is expected to aid the Conecuh-Sepulga Clean Water Partnership (CWP) in securing funds to work with the U.S. Fish and Wildlife Service to educate local landowners and farmers on proper land management techniques to help prevent further sedimentation of local rivers and streams. Locating the Gulf sturgeon's summer holding area locations within Conecuh/Escambia River will aid efforts to properly manage the designated critical habitat within this river. The objectives of this study were to track Gulf sturgeon movements and to identify summer holding areas within the Conecuh/Escambia River.

Study Area

Although the Conecuh/Escambia River is a single river body, this river shares two names. The river reach within the state of Alabama is referred to as the Conecuh River, and the reach of river located within Florida is named the Escambia River (Figure 1). The entire length of the Conecuh/Escambia River totals 362 km. Lakes Gantt and Point-A are two reservoirs that occur within the Conecuh/Escambia River system. Point-A Dam is the most downstream impediment of this river system, and the tailrace of this hydroelectric dam is located at river kilometer (rkm) 222 from the mouth of the Escambia River. The study area encompasses 132 km of the Conecuh River, from Point-A Dam to the Florida state boundary, and all 90 km of the Escambia River, from the Alabama state boundary to Escambia Bay, near Pensacola, Florida. All 222 km of the study area is designated as Gulf sturgeon critical habitat, and 38.6 km of this river reach is identified as potential Gulf sturgeon spawning habitat (USFWS, 2002).

The Sepulga River flows into the Conecuh River about 55.5 km downstream of Point-A Dam. A 25 km reach of this major Conecuh River tributary, from Bull Slough Landing in Paul, Alabama, to the Conecuh and Sepulga Rivers confluence, was included within the study area. The entire lengths of the Delaney and White Rivers, which are two major tributaries of the lower Escambia River, were also included within the study area. The Delaney River is 6.5 km long, and enters the lower Escambia River at rkm 29.4. The White River begins at a split from the Escambia River near rkm 17, and flows for 14.4 km before returning to the Escambia River just upstream of rkm five of the Escambia River.

Current land use within the Conecuh/Escambia River watershed is predominantly undeveloped lands with agriculture and silviculture uses. Threats to this river include the introduction of sediment and other non-point source pollutants as a result of these land uses and the many unpaved road crossings within this watershed.

METHODS

Study population collection occurred from 28 March 2005 to 16 May 2005 at Lora Point of Escambia Bay and at the Escambia River mouth (Figure 2). Gill nets with dimensions of 45.7 m long and 2.4 to 3.1 m deep, with 7.7 to 10.1 cm bar mesh, were used during the collection. Lora Point is located 3.45 km southwest of the Escambia River mouth, and was established as a sampling site for this study after learning from a local commercial fisherman that this location is a historical Gulf sturgeon fishing location. The eight netting attempts at Lora Point consisted of six to eight gill nets set at water depths of two to three meters. Two netting attempts occurred at the Escambia River mouth (Figure 2), and consisted of four gill nets placed in water depths of three to five meters and oriented toward the shoreline along the fringes of the navigational channel. The soak time of each gill net was limited to 1.5 hours to avoid asphyxiation of captured sturgeon. Netting was attempted within the

Conecuh and Escambia Rivers, but failed due to floating debris and numerous snags created by hurricanes Ivan and Frances that had recently passed through the study area.

Captured Gulf sturgeons were placed within an on-board live well system. Individuals were measured for total length (TL) and fork length (FL) to the nearest 0.1 cm, and were weighed with a spring scale to the nearest 0.1 kg. Each Gulf sturgeon received an external T-bar Floy tag (Floy FD-67) placed in both pectoral fins, and a passive integrative transponder (PIT) tag (Biomark) was injected into the base of the dorsal fin. Sub-adult (immature, 90 to 125 cm FL) and adult (mature, >125 cm FL) Gulf sturgeons (Sulak and Randall, 2004) were selected to receive an Advanced Telemetry Systems, Inc. (ATS) Model F1850 internal radio transmitter. The ATS transmitters operated at 40 MHz, measured 70 x 17 mm, weighed 26 g, had a battery life of 633 days at a pulse rate of 40 pulses-per-minute, and were equipped with a flexible external antenna. Radio transmitters were implanted into the abdomen of the selected Gulf sturgeons using the surgical procedure described by Fox et al. (2000). Released individuals were observed for proper orientation within the water and the ability to swim away unassisted.

Semi-monthly manual radio telemetry tracking within the Conecuh/ Escambia River began via boat on 27 April 2005; thirty days after the first Gulf sturgeon was radio-tagged. The initial-location radio telemetry searches lasted three days each, and were narrowed to two days each once the Gulf sturgeons' respective summer holding areas were located. Initial-location radio telemetry tracking was aborted and two-day specific summer holding area radio telemetry tracking began on 14 June 2005. The specific summer holding area radio tracking occurred on a twice monthly interval. Two ATS radio receivers, Models 2100 and 2000, were onboard during each search. The ATS Model 2100 was programmed to detect all of the Gulf sturgeons that were radio-tagged for the purpose of this study.

The ATS Model 2000 radio receiver was programmed to receive all radio tag signals of Gulf sturgeons tagged in the Yellow River (Figure 1) by the USFWS (n = 10) during the fall of 2004; in an attempt to document the migration of these individuals into the Conecuh/Escambia River. A 2.74 m whip antenna was attached to the bow of the tracking boat and was connected to the radio receivers by a t-splitter with two coaxial cables. The effective range of reception with this type antenna and transmitter was about 180 m. Once a radio-tagged Gulf sturgeon was detected, the 2.74 m whip antenna was removed from the radio receiver and replaced with a short-range antenna (Hightower et al., 2002).

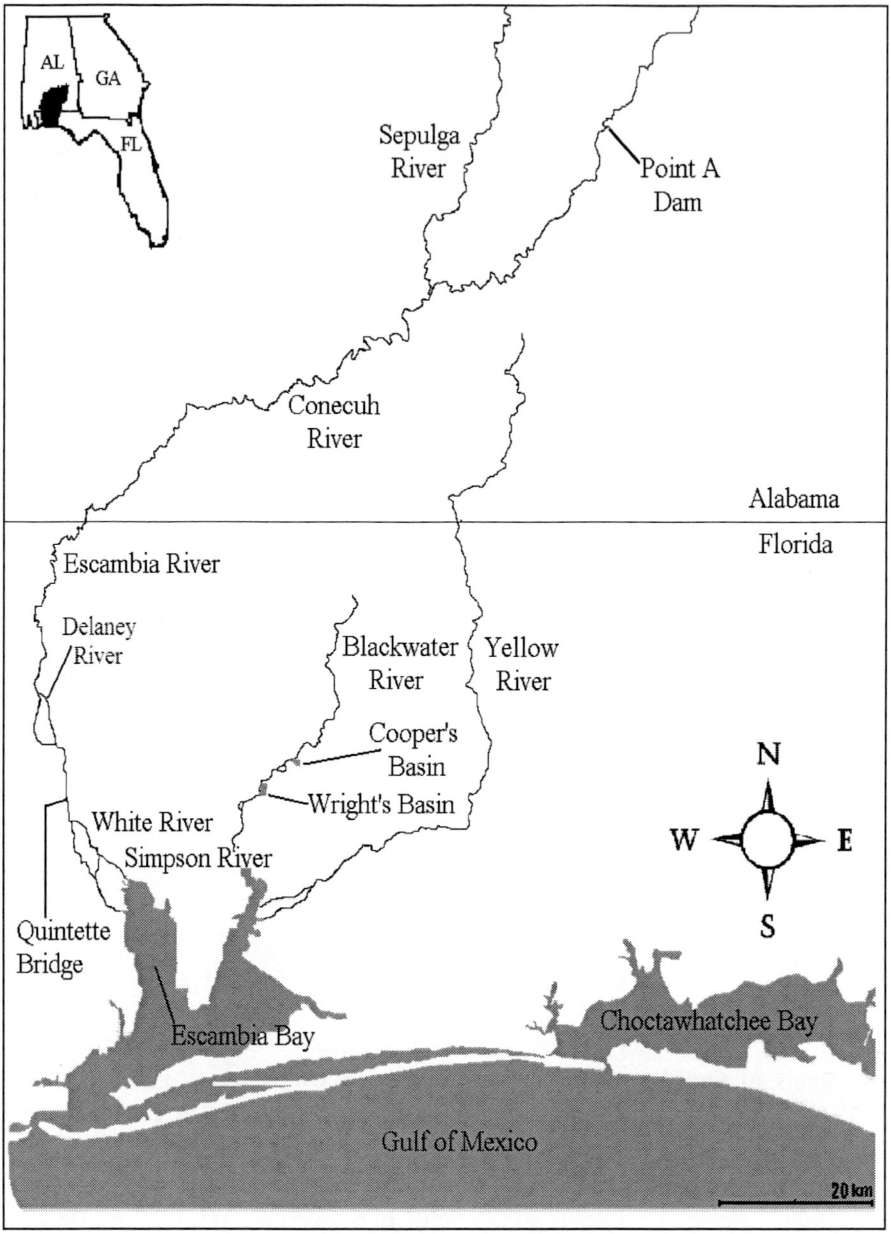

Figure 1. Map of the Conecuh/Escambia River system, from Point-A Dam in Andalusia, Alabama, to Escambia Bay in Pensacola, Florida.

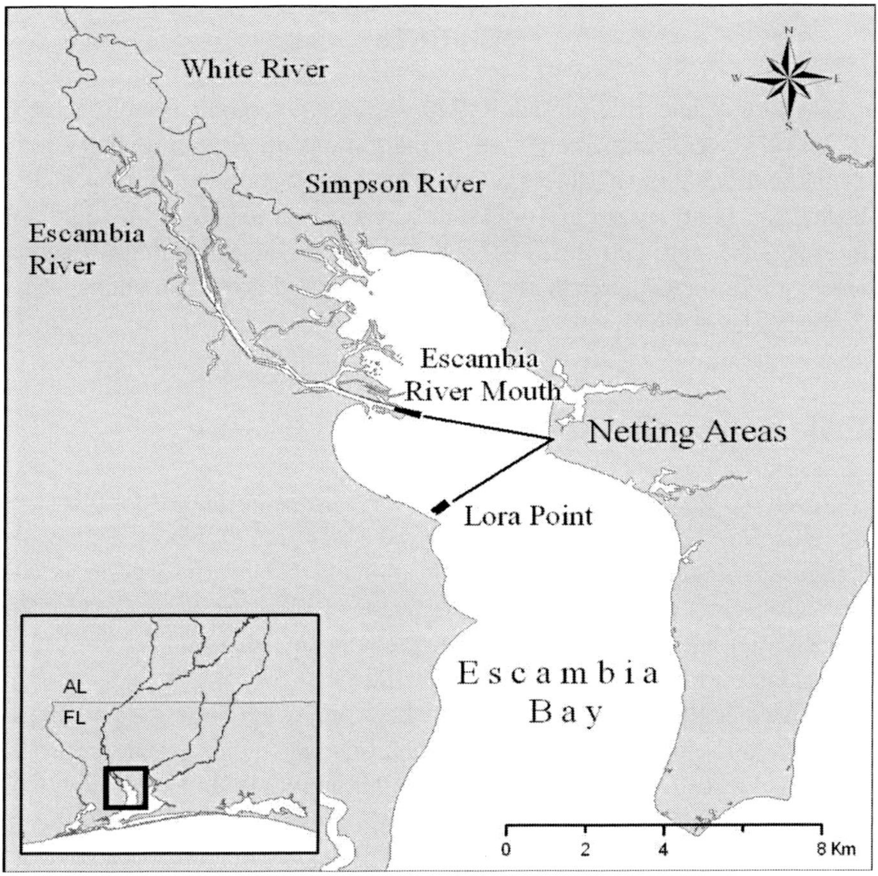

Figure 2. Map of the Escambia River and Escambia Bay Gulf sturgeon netting locations.

This method shortened the transmitter detection range, which allowed the position of the radio-tagged individual to be more precisely located. The latitude and longitude of each radio-tagged Gulf sturgeon telemetry relocation was recorded with either a Garmin GPS Map 76 or a Magellan Meridian Gold model handheld global positioning systems (GPS). Summer holding area lengths per Gulf sturgeon, which were determined by recording the most upstream and downstream radio detections of each individual within its chosen holding area, were recorded using the aforementioned GPS units. Flow volumes for the Escambia River were recorded from the U.S. Geological Survey (USGS) remote station # 02376033 near Molino, Florida. All tracking attempts ended on 2 December 2005.

RESULTS

Seven adult and five sub-adult Gulf sturgeons, ranging from 103 cm to 202 cm TL, were implanted with a radio transmitter during the netting attempts at Lora Point (Table 1). One of the adult individuals (Tag # 372) collected and radio-tagged at Lora Point was a gravid female. Two adult Gulf sturgeons (165 and 200 cm TL) collected at the Escambia River mouth received a radio transmitter implant (Table 1). One of these two adults, Tag # 912, was a gravid female.

Initial-Location Gulf Sturgeon Location Telemetry

Eleven of the fourteen Gulf sturgeons that were radio-tagged for this study were relocated via radio receivers within the study area (Table 1). Two of the fourteen radio-tagged individuals (Tag #s 330 and 372) were found in an area between Wright's and Cooper's Basins of the Blackwater River by USFWS biologists that were monitoring telemetry equipped fish in this river system. The location of these two Gulf sturgeons within the

Blackwater River was 55 km from where they were collected and tagged at Lora Point (Figure 1). One Gulf sturgeon that was radio-tagged during this study, Tag # 851, was never located during this study (Table 1). An adult Gulf sturgeon that was radio-tagged by the USFWS in the Yellow River during the fall of 2004, Tag # 761, was detected 13 times within the Escambia and Delaney Rivers (Table 1). The addition of this radio-tagged Yellow River Gulf sturgeon increased the study population to 12 individuals that supplied radio telemetry tracking data within the Conecuh/Escambia River study area.

There were 186 radio detections in the Conecuh/Escambia River during the monitoring period; with a mean of 15.5 radio detections per fish. The first radio-tagged Gulf sturgeon (Tag # 251) was located on 27 April 2005, and was 161 km upstream of its capture location at the Escambia River mouth on 28 March 2005. On 4 May 2005, this individual was telemetry relocated 37 km upstream of its 27 April 2005, location. The 4 May 2005, telemetry relocation of this Gulf sturgeon at rkm 198 of the Conecuh River, was the most upstream position that any of the radio-tagged Gulf sturgeons were found during this study. Craft et al. (2000) documented the occurrence of Gulf sturgeon spawning within the Conecuh River with the collection of eggs at five locations between rkm 161 and 170 of this river.

Table 1. Characteristics of radio-tagged Conecuh/Escambia River Gulf sturgeons

Date of capture	Capture location	Radio tag ID	Age class	Fork length (cm)	Total length (cm)	Weight (kg)	Number of radio tag detections
09/03/04	Yellow River	761*	Adult	No Data	No Data	49.5	13
03/28/05	Escambia River Mouth	251	Adult	147.3	165.1	32.8	20
03/28/05	Escambia River Mouth	912	Adult	189.2	200.0	61.7	17
03/29/05	Lora Point	562	Adult	162.6	170.2	36.6	5
03/29/05	Lora Point	191	Sub-adult	111.8	121.3	11.6	21
03/29/05	Lora Point	521	Adult	127.0	139.7	17.8	19
03/29/05	Lora Point	233	Adult	137.2	154.9	22.5	15
04/04/05	Lora Point	572	Sub-adult	111.1	127.0	11.2	18
04/04/05	Lora Point	851	Adult	178.4	189.2	54.5	0
04/13/05	Lora Point	291	Adult	182.3	201.9	59.0	16
04/14/05	Lora Point	372	Adult	158.8	190.5	51.9	1
04/27/05	Lora Point	592	Sub-adult	92.7	102.9	6.2	19
04/27/05	Lora Point	330	Sub-adult	108.0	120.0	19.1	1
05/04/05	Lora Point	101	Adult	146.1	162.6	32.0	15
05/16/05	Lora Point	641	Sub-adult	99.1	114.3	7.0	8

Note: asterisk refers to fish collected and radio-tagged by USFWS on Yellow River in 2004.

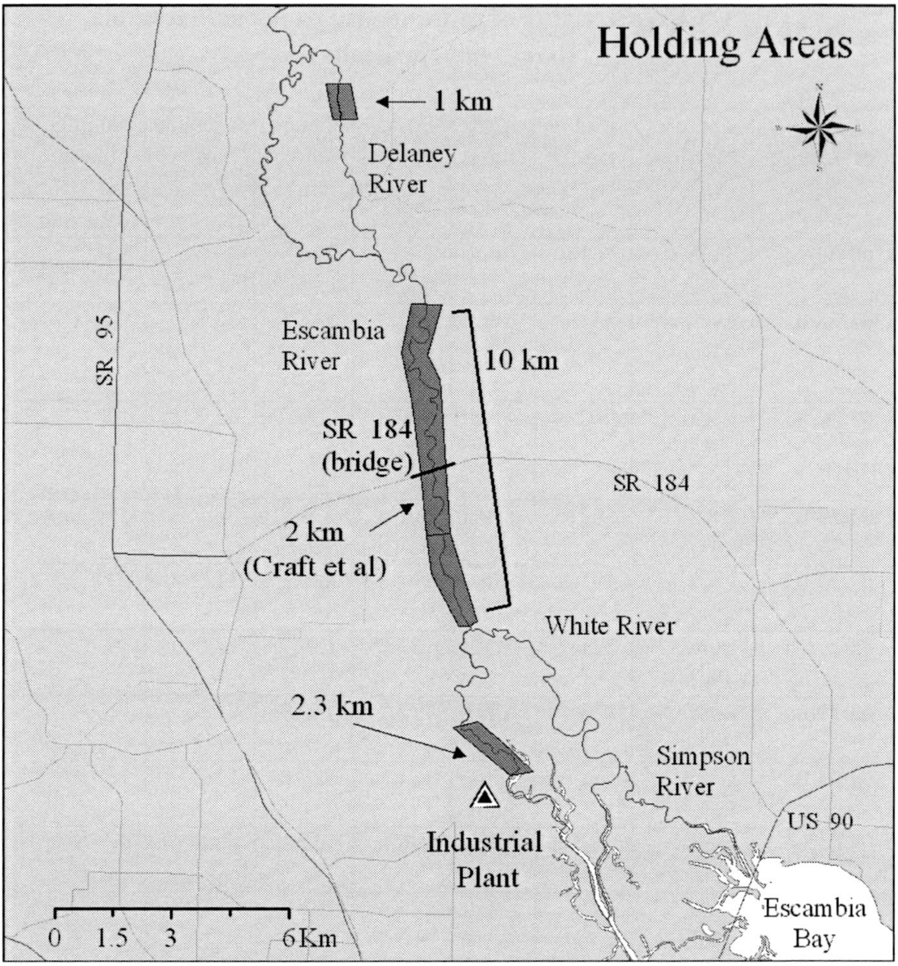

Figure 3. Map of the one, 2.3 and 10 kilometer Gulf sturgeon summer holding areas identified during this study within the Escambia River, Florida.

It is assumed that this individual (Tag # 251) continued upstream to complete its spawning run within the Conecuh River spawning grounds, but was not detected within this more upriver reach due to a one week lapse between radio tracking attempts. Tag # 251 was next detected 155 km downstream of its previous location, within the Escambia River, on 14 June 2005, and remained within this river reach for the remainder of the summer.

No other radio-tagged Gulf sturgeons were detected within the Conecuh River reach during five telemetry tracking trips that occurred April through

June 2005. No radio-tagged Gulf sturgeons were detected in the four tracking surveys conducted in the Sepulga River during the April through June 2005 time period. Radio-tagged Gulf sturgeons were documented using the White River for migratory activity during the initial-location radio tracking of this study. This is the first documentation of Gulf sturgeons using the White River as a migratory thoroughfare.

Summer Holding Area Telemetry and Gulf Sturgeon Movement Patterns

There were a total of 179 summer holding area radio telemetry detections, with an average of 14.9 detections per fish, among the 12 radio-tagged Gulf sturgeons of this study. The Conecuh and White Rivers radio tag relocations (n = 7) were deleted because the individuals were transient at the time of detection. The mean length of the summer holding areas used by individual Gulf sturgeons equaled 3.61 km (n = 12, 95% CI: 2.01-5.21), which was calculated by averaging the most upstream and most downstream locations that individual radio-tagged Gulf sturgeons were detected.

The summer holding area radio tracking efforts of this study documented 10.3 km of new Gulf sturgeon summer holding area habitat within the Escambia River, and confirmed two kilometers of summer holding area habitat that Craft et al. (2001) documented within this river during their study (Figure 3). Craft et al. (2001) noted that their Escambia River summer holding reach began at the Florida State Route (SR) 184 bridge (rkm 21), and stretched downstream for two kilometers. Our study documented that the Gulf sturgeon summer holding area at this location actually begins near rkm 27, and reaches downstream for 10 km. This study located another potential summer holding area 4.3 km downstream of the larger 10 km long Escambia River summer habitat (Figure 3). One radio-tagged Gulf sturgeon (Tag # 641) was detected within this suggested summer holding area that began at rkm 10.4 of the Escambia River, and stretched upstream for 2.3 km.

The Yellow River Gulf sturgeon (Tag # 761) that entered the Escambia River during this study migrated to a reach of the Delaney River for its summer refuge. This radio-tagged individual was telemetry relocated in the Delaney River on eight separate occasions from 13 July through 19 October 2005, and occupied a summer holding area that began 4.3 km upstream of the Delaney-Escambia River confluence, and reached upstream for 1 km (Figure

3). No summer holding areas were observed in the Conecuh, Sepulga or White rivers.

DISCUSSION

No summer holding areas were noted within the Conecuh, Sepulga or White rivers during this study. A 10 km and a 2.3 km summer holding area were located and documented within the Escambia River, and another was found within the Delaney River. Although Tag # 641 was the only radio-tagged Gulf sturgeon documented using the 2.3 km-long summer holding area in the lower Escambia River, it is possible that other Gulf sturgeons were also holding within this same area. Due to the confirmation of Tag # 641 and the probability of other non-telemetered Gulf sturgeons using this river reach as summer holding habitat, it is recommended that this area be included within the overall amount of summer holding area habitat documented by this study. One radio-tagged adult Gulf sturgeon (Tag # 761) was detected on nine occasions, from 13 July through 19 October 2005, within the Delaney River. It is recommended that this one kilometer reach of the Delany River also be added to the summer holding area habitat designations of the Conecuh/Escambia river system due to the documentation of Tag # 761 using this river reach as summer habitat, and because of the numerous observations of breaching Gulf sturgeons within this location during the tracking efforts. This is the first documentation of Gulf sturgeons using the Delaney River for summer holding habitat.

We received anecdotal information from fishermen and from personnel of an industrial facility that Gulf sturgeons were often observed breaching during the summer and early autumn in an area of the Conecuh River associated with the confluence of Murder Creek, near Brewton, Alabama (Figure 1). Investigations by the USFWS within this area during this study did not produce evidence of Gulf sturgeons residing within this river reach. However, Craft et al. (2001) observed several Gulf sturgeons at this site during the summer of 2001, and personnel of the USFWS and the Alabama Department of Conservation and Natural Resources (ADCNR) collected and tagged 39 Gulf sturgeons at this location in 2002 (USFWS, 2002). The USFWS and ADCNR netting efforts occurred during a low water event of the Conecuh River; within a depression of the river bottom located at the Murder Creek – Conecuh River confluence. This depression could serve as a congregational area for Gulf sturgeons using the Conecuh River as summer holding area habitat during periods of low river flow. Our netting efforts at this location occurred during a prolonged high water event of the Conecuh River, and this may explain our lack of netting success due to the abundant availability of other Conecuh River areas that Gulf sturgeons could use as summer holding area habitat during prolonged high water events. Although none of the Gulf sturgeons radio-tagged for this study was detected using this Conecuh River reach as a summer holding area, the information gathered by various agencies suggests that this site may be used as a summer holding area contingent upon water flows. Further summer investigations at different water flow volumes are needed to better define the significance of this Conecuh River site.

Movement and Habitat Selection

Rogillio et al. (2007) reported that radio tagged sub adult and adult Gulf sturgeons were spatially separated in their summer habitat within the Bogue Chitto River, Louisiana. Our findings did not indicate any separation of summer habitat use between radio-tagged sub adult and adult Gulf sturgeons, and we also observed that the sites chosen by radio-tagged Gulf sturgeons for summer holding areas did not visually differ from surrounding areas of the river. However, this study's radio-tagged Gulf sturgeon population preferred to remain within their designated summer holding area. Only one of the 179 summer holding area radio detections (0.56%) occurred outside of the defined summer holding reach during the tracking efforts. This finding supports that of Zehfuss et al. (1999), who noted that 75% of their radio-telemetered Gulf sturgeon detections were within the chosen summer holding area of the Apalachicola River.

On 10 August 2005, this wandering radio-tagged individual (Tag # 233) was detected within the White River; one kilometer downstream of where this river splits from the Escambia River. The distance that this Gulf sturgeon was detected outside of the Escambia River summer holding area is consistent with the observations of Carr et al. (1996) in the Suwannee River. They noted that Gulf sturgeon movements were limited to two kilometers upstream or downstream of their selected summer holding area within the Suwannee River. Chapman and Carr (1995) also observed Gulf sturgeons remaining within one kilometer upstream or downstream of their summer holding areas in the Suwannee River. Tag # 233 was next detected back within its summer holding area near the SR-184 bridge on 7 September 2005, and was never again detected outside of the summer holding area bounds. Our observation agreed with the Apalachicola River study of Zehfuss et al. (1999), where their Gulf sturgeons also returned to the summer holding area after venturing beyond the holding area boundaries.

Though these summer holding area data suggest that Gulf sturgeons rarely venture beyond the bounds of a chosen summer holding area, most of the data were obtained using radio-telemetry equipment with manual tracking techniques. This method limits availability of movement observations due to the lapse of time that can occur between manual tracking efforts. Another limiting factor is that Gulf sturgeons often display more nocturnal mobility, and all of our observations occurred during the day. We recommend that future studies of the spawning migration and summer holding areas of Conecuh/Escambia River Gulf sturgeons use either sonic or radio transmitters with remote data-logging receivers. The use of these technologies can permit continuous tracking of Gulf sturgeon movements, which will supply more detailed data of the desired behaviors.

The radio-tagged sub adult Gulf sturgeon (Tag # 641) that selected its summer holding area at rkm 10.4 of the Escambia River exhibited an interesting behavioral pattern. This individual was detected on 7 September 2005, holding near the barge canal of an industrial facility (Figure 3). The area was free of barges at the time of observation, and the Gulf sturgeon was located in 4.88 m of water. This Gulf sturgeon was observed the next day 1.77 km upstream from its 7 September 2005 location. Upon traveling downstream to the previous location of this individual, there were two large barges moored where this Gulf sturgeon was located the day before. The river at this location was about 37 m wide, and was just large enough for one barge to moor along one side of the river and still allow vessel traffic to navigate the river. It is suspected that the disturbance from the barges' movement in such a narrow channel may have prompted this Gulf sturgeon to leave its location. This individual was never again observed in its original location at the industrial facility, and remained in a 1.19 km stretch of river (n = 5 observations), 0.7 km upstream of the facility. Although this observation only involved one fish, further study of sudden disturbances triggering an avoidance behavior of Gulf sturgeons within summer holding habitats may help further the understanding of summer refugia selection and behavior.

Periods of high flow were common on the Escambia River during the summer of 2005. The observed mean daily flow during the study period was nearly twice as high as the 20-year mean flow record for the USGS gauge # 02376033 near Molino, Florida (Figure 4). This study observed that prolonged periods of elevated flow volumes within the Escambia River during the Gulf sturgeon summer holding months did not appear to affect radio-tagged Gulf sturgeons' fidelity toward their selected summer holding area or their migratory behavior. This observation was consistent with those of Wooley and Crateau (1985) and Zehfuss (2000) in the Apalachicola River, Foster and Clugston (1997) in the Suwannee River, and Hightower et al. (2002) in the Choctawhatchee River.

Spring inflows were often closely associated with congregating Gulf sturgeons in the Suwannee and Choctawhatchee Rivers (Chapman and Carr, 1995; Carr et al., 1996; Foster and Clugston, 1997; Hightower et al., 2002). This was not the case in the Escambia and Delaney River summer holding areas. There were only two documented springs in the Escambia River, yet the Florida reach of the Choctawhatchee River had seven springs within or near its main channel, and the Suwannee River had over 200 springs located within or near its main stem (Katz et al., 2003; Spechler and Schiffer, 1995). The nearest spring inflow to an Escambia River summer holding area was located 13.2 km upstream of the summer habitat located near the SR-184 bridge. Foster and Clugston (1997) pointed out that Gulf sturgeons in the Suwannee River could have been observed congregating near spring inflows simply because of the numerous springs located within this river's drainage, which our findings support.

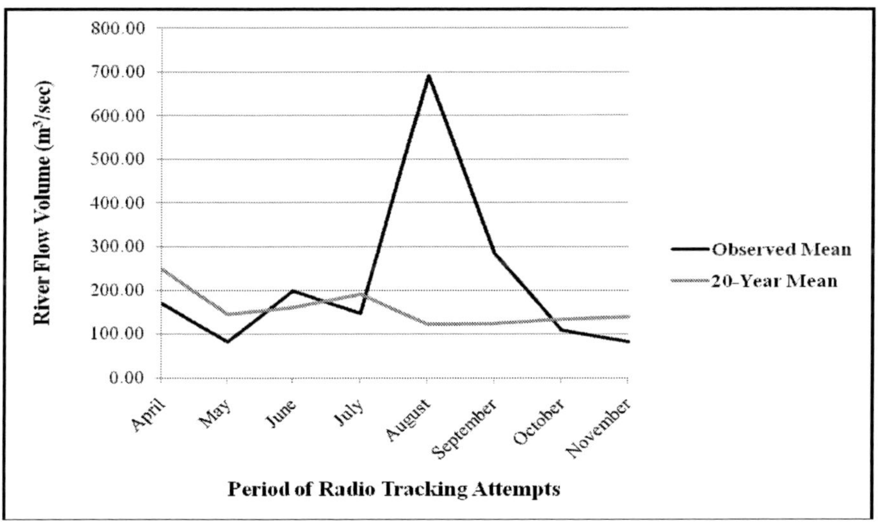

Figure 4. Mean Escambia River flow volumes observed April through November 2005 compared with 20-year mean flow volumes at the U.S. Geological Survey remote station # 02376033, near Molino, Florida, during radio tracking attempts.

Several Escambia River Gulf sturgeons were telemetry detected within scour holes and on the leeward side of ledges along the river's bottom. This behavior was well documented by Sulak and Clugston (1999) in the Suwannee River. Moser and Ross (1995) also observed this behavior in Atlantic sturgeons within the Lower Cape Fear River, North Carolina. It was originally suggested that these holes might serve as a form of temperature refuge, but Gulf sturgeons most likely use these depressed areas to escape strong river currents; thereby conserving energy during their long period of fasting (Sulak and Clugston, 1999).

CONCLUSION

The overall goal of this study was to locate and document Gulf sturgeon summer holding area habitats within the Conecuh/Escambia River Basin. This study confirmed the two-kilometer summer holding area documented by Craft et al (2001), and expanded the overall amount of suitable Gulf sturgeon summer holding area to 12.3 km within the Escambia River. A one kilometer long summer holding area was located and documented within the Delaney River. This study also tracked and documented a mature Gulf sturgeon traveling far upstream into the spawning habitat of the Conecuh River. The White River was documented for the first time as a thoroughfare for migrating Gulf sturgeons. Summer holding area movements of Gulf sturgeons within the Escambia River were consistent with the findings of other researchers, and we observed Gulf sturgeons migrating into and out of the river at water temperatures similar to those previously documented throughout the Gulf sturgeon's range. Unlike observations within the Choctawhatchee and Suwannee Rivers, the Escambia River Gulf sturgeons did not prefer locations downstream of spring inflows. Many Gulf sturgeons in the study population used scour holes and the leeward side of ledges, which may explain why they did not appear to be influenced to move downstream by the abnormally high flows of the river during the study period.

Documentation and protection of Gulf sturgeon summer holding areas are essential for the Gulf sturgeon's successful recovery. Measures must be taken to insure that water quality in the Conecuh/Escambia River system is maintained at standards that can support a healthy aquatic community. Environmental disturbances that may pose a threat to disrupting Gulf sturgeon summer holding areas within the Conecuh/Escambia River system also need to be documented and rectified. Further investigations using sonic or radio transmitters and passive receivers are needed to determine whether additional Gulf sturgeon summer holding habitat is present within this riverine system.

ACKNOWLEDGMENTS

We thank PowerSouth, Inc., for project funding. Bob Jarvis and Alicia Newberry of the USFWS Panama City Field Office; Keith Stephens, Larry Spann, Melissa Mathews, and Lisa Wiggins of PowerSouth, Inc., and Michelle Duncan of the National Marine Fisheries Service were invaluable during both field and office work.

REFERENCES

Carr, S.H., Tatman, F. & Chapman, F.A. 1996. Observations on the natural history of the Gulf of Mexico sturgeon (*Acipenser oxyrinchus desotoi* Vladykov 1955) in the Suwannee River, southeastern United States. *Ecology of Freshwater Fish* 5:169-174.

Chapman, F.A. & Carr, S.H. 1995. Implications of early life stages in the natural history of the Gulf of Mexico sturgeon, *Acipenser oxyrinchus desotoi*. *Environmental Biology of Fishes* 43:407-413.

Craft, N.M., Russell, B. & Travis, S. 2001. Identification of Gulf sturgeon spawning habitats and migratory patterns in the Yellow and Escambia River systems. *Final Report to the Florida Marine Research Institute,* Fish and Wildlife Conservation Commission, Tallahassee, Florida.

Foster, A.M. & Clugston, J.P. 1997. Seasonal migration of Gulf sturgeon in the Suwannee River, Florida. *Transactions of the American Fisheries Society* 126:303-308.

Fox, D.A., Hightower, J.E. & Parauka, F.M. 2000. Gulf sturgeon spawning migration and habitat in the Choctawhatchee River system, Alabama – Florida. *Transactions of the American Fisheries Society* 129:811-826.

Hightower, J.E., Zehfuss, K.P., Fox, D.A. & Parauka, F.M. 2002. Summer habitat use by Gulf sturgeon in the Choctawhatchee River, Florida. *Journal of Applied Ichthyology* 18:595-600.

Huff, J.A. 1975. Life history of the Gulf of Mexico sturgeon, *Acipenser oxyrhynchus desotoi,* in Suwanne River, Florida. Florida Department of Natural Resources, Marine Research Laboratory Publication 16. 32 pp.

Katz, B.G., Griffin, D.W., Swarzenski, P.W., Walsh, S.J. & Jelks, H.L. 2003. Florida springs interdisciplinary science study. U.S. Geological Survey, Publication FS-008-03.

Mason, W.T. & Clugston, J.P., Jr. 1993. Foods of the Gulf sturgeon in the Suwannee River, Florida. *Transactions of the American Fisheries Society* 122:378-385.

Moser, M.L. & Ross, S.W. 1995. Habitat use and movements of shortnose and Atlantic sturgeons in the Lower Cape Fear River, North Carolina. *Transactions of the American Fisheries Society* 124:225-234.

Rogillio, H.E., Ruth, R.T., Behrens, E.H., Doolittle, C.N., Granger, W.J. & Kirk, J.P. 2007. Gulf sturgeon movements in the Pearl River drainage and the Mississippi Sound. *North American Journal of Fisheries Management* 27:89-95.

Smith, T.I. 1990. *Culture of North American sturgeons for fishery enhancement.* National Oceanic and Atmospheric Administration Technical Report National Marine Fisheries Service 85, Kyoto, Japan.

Spechler, R.M. & Schiffer, D.M. 1995. Springs in Florida. U.S. Geological Survey, Publication FS-151-95.

Sulak, K.J. & Clugston, J.P. 1999. Recent advances in life history of Gulf of Mexico sturgeon, *Acipenser oxyrinchus desotoi*, in the Suwannee River, Florida, USA: a synopsis. *Journal of Applied Ichthyology* 15:116-128.

Sulak, K.J. & Randall, M. 2004. *Standardization of Gulf of Mexico sturgeon life stage terminology for conservation biology and population management use.* Gulf Sturgeon Research Workshop, Mandeville, Louisiana.

U.S. Fish and Wildlife Service. 1991. Endangered and threatened wildlife and plants; determination of threatened status for the Gulf sturgeon. *Federal Register* 56(189):49653-49658.

U.S. Fish and Wildlife Service. 2002. *2002 Annual Report.* Panama City Fisheries Resource Office, Florida. 34 pp.

U.S. Fish and Wildlife Service and Gulf States Marine Fisheries Commission. 1995. *Gulf Sturgeon Recovery Plan.* U.S. Fish and Wildlife Service, Atlanta, Georgia.

U.S. Fish and Wildlife Service and National Oceanic and Atmospheric Administration. 2003. Endangered and threatened wildlife and plants; designation of critical habitat for the Gulf sturgeon; final rule. *Federal Register* 68:13370-13495.

Wooley, C.M. & Crateau, E.J. 1985. Movement, microhabitat, exploitation, and management of Gulf of Mexico sturgeon, Apalachicola River, Florida. *North American Journal of Fisheries Management* 5:590-605.

Zehfuss, K.P., Hightower, J.E. & Pollock, K.H. 1999. Abundance of Gulf sturgeon in the Appalachicola River, Florida. *Transactions of the American Fisheries Society* 128:130-143.

Zehfuss, K.P. 2000. *The status, movement, habitat preference, and monitoring of Gulf sturgeon in several Florida rivers*. Ph.D Dissertation. North Carolina State University, Raleigh, NC.

In: Watersheds
Editors: G. L. Kuhn and J. R. Emery
ISBN 978-1-62081-057-6

Chapter 4

CLIMATE CHANGE AND ITS EFFECT ON FLORIDA WATERSHEDS

Smita Barkataky[1] and Debashish Goswami[*,2]

[1]Biological Scientist, Southwest Florida Research and Education Center, University of Florida, Immokalee, Florida, US

[2]Post-Doctoral Research Associate, Southwest Florida Research and Education Center, University of Florida, Immokalee, Florida, US

ABSTRACT

While there are a number of possible causes including greenhouse gases, solar radiation, land use changes that may contribute to the changes in the worldwide climate, the reality is that regardless of the causes, much of the scientific community is in agreement that climate change is real. The global climate change during the recent years has been considered as one of the biggest threats to Florida. Many factors including solar radiation, atmospheric composition, cloud cover, ocean currents, and land activities influence the climate and make long-term changes. Florida's topography and its warm climate make the state vulnerable to the effects of climate change. Although Florida has abundant sources of surface water and groundwater, it is reported that water levels in Florida have dropped at an average rate of 0.10 to 0.15 m

* Corresponding author: Dr. Debashish Goswami, Biological Scientist, Southwest Florida Research & Education Center, University of Florida, 2685 State Road 29 North, Immokalee, Florida 34142, Telephone number: (239) 565- 3352, Fax number: E-mail address: debashishgoswami@ hotmail.com.

per year since 1950. This review article examines the impact of global climate change on Florida, especially on its watersheds, and potential climate change adaptation strategies.

INTRODUCTION

The Earth's climate has changed throughout history. Many factors which include solar radiation, atmospheric composition, cloud cover, ocean currents, and land activities (e.g. deforestation and urbanization) influence climate and make long-term changes (SFWMD, 2009). Before the industrial revolution (1820-1870), natural causes such as volcanic eruptions, changes in the Earth's orbit, and the amount of energy released from the sun were the primary factors affecting the Earth's climate (USEPA, 2009). Since the beginning of the industrial revolution, human activities have led to increased concentration of greenhouse gases (GHG) in the atmosphere (Smith and Tirpak, 1989). One of the primary reasons for the climate change is reported as increased surface air temperature resulting in decreased ice mass (Solomon et al., 2007). Studies indicated that the changes in climate are, to a great extent, caused by human activities (NRC, 2010) and are generally not reversible (Solomon et al., 2009). However, climate change and its variability at regional level are not completely understood yet (SFWMD, 2009).

Over this century, anthropogenic climate change will cause significant challenges to Florida in particular and to the world in general, as it will affect nearly every aspect of our planet's living systems and human economy (Mulkey et al., 2007). The task of predicting Florida's climate scenario over the next 100 years is challenging, given the current state of climatology (Mulkey et al., 2007). Florida has 52 large watersheds or basins. In order to best protect and restore them, Florida Department of Environmental Protection (FDEP) has organized these watersheds into 29 groups of basins to make environmental management easier, more effective and more uniform across programs (FDEP, 2011). Florida Department of Environmental Protection is involved in managing the quality and quantity of water through its relationship with the state's five water management districts: Northwest Florida Water Management District, Suwannee River Water Management District, St. Johns River Water Management District, South Florida Water Management District, and Southwest Florida Water Management District. In addition to the Water Management Districts, several Federal and State organizations are involved in protecting and restoring the state's water resources from the Keys to the

Panhandle of Florida. However, the global climate change during the recent years has been considered as one of biggest threats to the Florida watersheds (Bloetscher et al., 2011). With nearly 1,920 km of coastline and 95% of its population living within 56 km of the coast, Florida is uniquely susceptible to climate change (WWF, 2012).

Although Florida has abundant sources of surface and groundwater, it is reported that water levels in Florida have dropped at an average rate of 0.10 to 0.15 m per year since 1950 (Barnett, 2007). Continuous land use conversion triggered by the increasing population in Florida, which has more than doubled during the last three decades (Karl et al., 2009), has brought many challenges for the future generation with respect to heat stress, water scarcity, sea-level rise and extreme weather events (Karl et al., 2009). In fact, scientists have already observed changes in Florida consistent with the early effects of global warming: retreating and eroding shorelines, dying coral reefs, saltwater intrusion into inland freshwater aquifers, and warmer air and sea-surface temperatures (Fiedler et al., 2001). This review article examines the impact of global climate change on Florida especially on its watersheds, and potential climate change adaptation strategies.

EFFECTS OF CLIMATE CHANGE

Sea Level Rise

To predict the sea-level rise correctly in a warming world is a challenging task. According to the projections for the twenty-first century, the 145 million people living within a meter of the present sea level risk losing their land and homes due to possible sea level rise (Clark et al., 2009). Florida's low elevation makes the state highly susceptible to sea-level rise; and its current exposure to hurricanes and warm climate make the state vulnerable to the consequences of climate change (Fu¨ssel, 2007). Several studies have been conducted all over the world to study sea-level rise (Gornitz and Lebedeff, 1987; Maul, 1993; Hansen, 2007; Rahmstorf, 2007; Martin et al., 2011; Rignot et al., 2011). Global average sea level has risen at 3.1 ± 0.70 mm/year since 1993 (Cazenave et al, 2008) with contributions from thermal expansion and melting of glaciers and ice sheets (Bloetscher et al., 2011). Florida may experience sea level rise in the order of 12.7 to 50.8 cm within the next 50 years (SFWMD, 2009). This high rate of sea level rise has the potential to threaten low-lying coastal areas of Florida. Zhang (2011) estimated that only a

0.5 m sea-level rise would inundate a large area of the Everglades National Park (ENP) and the marsh areas in southeastern Miami-Dade County of Florida. Accelerated sea level rise is likely to be accompanied by increasing temperatures (IPCC, 2007) and more intense tropical storms and hurricanes (Webster et al., 2005). Sea-level rise could cause erosion, saltwater intrusion, inundation, and increased flooding along the coastal zones of Florida (Zhang, 2011). Among the several harmful effects of sea level rise, salt water intrusion is the greatest threat to Florida's watersheds. By the late 1990s, saltwater intrusion extended approximately 30 km inland in the Shark River Slough region of ENP, and about 10 and 15 km inland in the regions of C-111 Basin and Taylor Slough in Florida, respectively (Fitterman et al., 1999). The boundary between fresh and saline groundwater in this region was reported to be diffuse, contrary to a sharp transition (Price et al. 2006). Reports indicated that the Biscayne aquifer in Florida could become salty due to intrusion of salty sea water with only a 92 cm rise in sea level (Scheraga and Grambsch, 1998). Sea level varies across the globe as a result of ocean circulation patterns, regional gravitation differences, deformation of the Earth's surface, and shifts in its rotation axis in response to ongoing changes in loading by glaciers and ice sheets, atmospheric pressure and regional salinity changes (IPCC, 2007). The maximum sea level rise due to the collapse of the Antarctic ice sheet is concentrated along the Pacific and Atlantic seaboard of the United States, where sea level rise is about 25% greater than the global mean (Bamber et al., 2009). Southern Florida is especially vulnerable given the low relief and gradual rise in elevation throughout the system (Pearlstine et al., 2010).

Temperature

Intergovernmental Panel on Climate Change (IPCC) examined trends and variability among climate-simulation models from 18 modeling centers around the world (IPCC, 2007). They estimated changes in temperature, precipitation, storms, ocean circulation, and sea level that are expected to occur over the next 100 years. Warming trends are being observed in Florida (Soule, 2005). Estimates based on simulations of GHG emission scenarios indicated that by 2100, temperatures in southern Florida will be 2.0 to 2.5 °C higher than the 2000 annual average if emissions are moderate (IPCC, 2007) or 5.5 °C higher if emissions are higher (Stanton and Ackerman, 2007).

Increased concentrations of Carbon dioxide (CO_2) and GHGs result in increased atmospheric temperature. The atmospheric CO_2 concentration, which has increased from 280 parts per million (ppm) in pre-industrial time (Neftel et al., 1982; Friedli et al., 1986) to 380 ppm at present, is expected to reach 700 ppm or more towards the end of the 21st century (Solomon et al., 2007) as a result of the change in land-use and combustion of fossil fuel. Average global temperature has risen by 0.76°C since 1850 and is projected to increase another 0.5-1.0 °C due to GHGs that have already been added to the atmosphere (Dawson and Spannagle, 2009). Average annual temperature of Florida during 1974-2001 increased approximately 0.17 °C compared to during 1895-1921; and the period 1990-2001 was found to be exceptionally warm (Winsberg, 2003). The same study also reported that Florida's hottest years during 1895-2001 were 1990 and 1998. Six of the top 27 hottest summers and five of the 27 hottest springs in Florida occurred between 1990 and 2001, while the State did not experience any coldest winter or summer between 1990 and 2001. According to Gaffen and Ross (1999), there is a warming of several tenths of a degree per decade over most of the United States, including Florida, in winter, spring, and summer seasons.

Stanton and Acemran (2007) reported that Florida's annual average temperature will be 2.8 ºC higher than today in 2050 and 5.6 ºC higher in 2100. Projected Florida temperatures will be higher compared to the global mean temperatures for the corresponding periods (Stanton and Acemran, 2007). Annual average temperature increase (in ^{o}C above year 2000 temperature): Florida = 1.3 and global mean = 1.2 (year 2025), Florida = 2.7 and global mean = 2.4 (year 2050), Florida = 4.1 and global mean = 3.6 (year 2075), and Florida = 5.4 and global mean = 4.8 (year 2100). SFWMD (2009) reported that considerable uncertainties exist in the regional projections of temperature resulting from climate change. Water demands are expected to increase significantly as temperatures rise. Agricultural and domestic water use will likely rise as temperatures go up. The potential for wildfires, especially during the dry season, will also increase. Even modest increases in temperature and rainfall reductions during the wet season (May-October) of the year could result in extended droughts and uncertain recharge in the Everglades wetland ecosystems that may produce dramatic changes to the Everglades (Pearlstine et al., 2010).

Precipitation

Many studies support the fact that global warming would lead to changes in annual or seasonal precipitation (Crisci et al., 2002). Groisman and

Easterling (1994) studied the annual precipitation and snowfall changes over Canada, Russia and USA, and the trend analysis confirmed earlier findings (Bradley et al. 1987; Diaz et al. 1989; Vinnikov et al. 1990) of possible increase in precipitation. Climate change affects the frequency and intensity of extreme events such as floods and droughts. This was supported by some greenhouse simulation studies in terms of changes in precipitation intensity (Noda and Tokioka, 1989; Mearns et al., 1990). A relatively large increase in summer precipitation was predicted in a climate change scenario study for Florida by Ewel and Smith (1992). This was supported by Mulholland et al. (1997), who believed that there was an increase in rainfall in coastal zones and in the southern regions of Florida. The increasing rainfall coupled with sea level rise, may result in increasing water depth and lengthening the hydroperiods of the wetlands (Mulholland et al., 1997) in the watersheds. In addition to that, increase in precipitation could result in increased runoff from agricultural watersheds, which may lead to high nutrient loadings and severe eutrophication of some lakes and wetlands (Ewel and Smith, 1992).

Winsberg (2003) observed that the rainfall pattern in Florida has become inconsistent during 1990-2001. During the study period, there were occurrences of both extreme wet and dry summers. Records were established for 27 wettest and 27 driest summers over a period of 107 years among which, five wettest and four driest summers occurred just in an 11-year period (1990-2001) (Winsberg, 2003). SFWMD (2009) quoted an IPCC report predicting up to 20% less rainfall for South Florida for the period between 2080 and 2099 in comparison to historical measurements from 1980 to 1999. The intensity of rainfall events has also been changing. There has been an increase in heavy downpours in many parts of the region, while the percentage of the region experiencing moderate to severe drought increased over the past three decades. Climate models predicted longer stretches of dry days between more intense rain events in Florida. The combination of these conditions increases the likelihood of more severe floods in the region. Water demands may increase if there is not enough rainfall to support water supply and irrigation needs. Without adequate water to maintain vegetation during droughts, the performance of Stormwater Treatment Areas (STAs) may be impacted when rains return (SFWMD, 2009)

Water Resources

In the second half of the twentieth century, Florida population grew from 2.8 to 16.0 million; and water use for public (non-agricultural) purposes increased by 1330% (Barnett, 2007). The United States Geological Survey (USGS) reported on the cumulative effects of enormous groundwater withdrawals from the Floridan aquifer between 1950 and 2000. Geologists Richard Marella and Marian Berndt found alarming groundwater-level declines and saltwater intrusion throughout the aquifer. In northeastern Florida, water levels have been dropping at an average 0.10 to 0.15 m per year since 1950. Along the coastal panhandle of Florida, water levels dropped 30.5 m since 1950 (Barnett, 2007). Various human activities such as the development and management of river flow could disable many of the natural mechanisms that allow the wetlands to adapt to a rising sea, and thereby substantially increase the loss of wetlands over what would occur naturally and in some areas the impacts could be so severe that entire ecosystems could be lost (Titus, 1988). Rising seas will likely have direct impacts on coastal beaches, infrastructure, and wetlands due to high tides and storm surges (SFWMD, 2009). A better understanding of the uncertainties in the sea level rise projections and a detailed analysis are needed to identify the impact of projected sea level rise on flooding, saltwater intrusion into freshwater aquifers, and wetlands in the coastal areas. With sea levels rising gradually, the productive and diverse estuarine zone would probably grow. This landscape would change as some wetlands become open water and some terrestrial habitats convert to mangroves (SFWMD, 2009). Stanton and Acemran (2007) reported that the Everglades would be severely impacted by sea level rise. Rising sea levels would cause water to encroach 19.2 to 38.4 km into the low lying area of the Everglades, leaving the lower Everglades completely under water. As large parts of the Everglades are converted into open water, nurseries and shelter for many fish and wildlife species will be lost (Stanton and Acemran, 2007). Bloetscher (2009) mentioned that much of the studies on climate change have focused on the Everglades ecosystem; however, the effects of climate change can be felt in other parts of Florida too. In northeast Florida, the area relies on the St. Johns River and the Floridan Aquifer. Rainfall variation will affect runoff in the St. Johns River also.

Mulhollanda et al. (1997) identified the major ecological effects of climate change on freshwaters of the southeastern part of the United States, including Florida as: an increase in the rates of primary production, organic matter decomposition and nutrient cycling as a result of temperature rise; reduction in habitat for water species, and water quality degradation as a result of lower groundwater flows; and increase in eutrophication of Florida lakes as a result of higher runoff rates from urban and agricultural areas. Fiedle et al. (2001) reported that anticipated sea level rise will very likely cause vast inundation and destruction of coastal wetlands and salt water intrusion into the low lying area of the Everglades watershed. This threatens much of the lower Everglades because these ecosystems cannot tolerate flooding of this extent. However, unless the causes of global warming are addressed, a major portion of even the healthiest Everglades ecosystem will be inundated and degraded.

The greatest risk to wetlands and other ecosystems originates from a possible decrease in precipitation and higher temperatures, especially if they occur over several consecutive years. Frequent and too many consecutive years of droughts are harmful for Florida watersheds. Droughts also greatly increase the competition for water supply for human uses, further reducing freshwater available to natural wetland and coastal ecosystems. On the other hand, high precipitation results in unnatural fluctuations in the large fresh water bodies. This affects the salinity and the biodiversity of the coastal estuaries because the excess freshwater is discharged to the estuaries in the east and west coasts of Florida (Fiedle et al., 2001). Climate change may result in less water being available to maintain groundwater supplies, provide baseflow to streams, and provide adequate soil moisture for cultivation in Florida and many other parts in the world (Erwin, 2009).

Land Use

The most important economic issue facing Florida during the years to come is probably land use. The forest, agricultural, and natural lands in Florida have yet to be managed for GHG offsets and mitigation. Florida's population will likely increase by at least 50% over the next twenty-five years, and may double in fifty years. Urban development, suburban growth, increased human population densities, and reduced agricultural and forest lands will be the inevitable result of this population increase unless growth is managed wisely for enhancing sustainability (Mulkey, 2006). The negative effects on land use and natural resources are made ever more critical by the unfolding consequences of climate change, which will impact densely populated coastal regions as sea level rises. Stanton and Acemran (2007) reported the amount of land area in the vulnerable zone in Florida by land use for a 0.69 m sea-level rise scenario as follows: agriculture (135 sq km), developed (1,126 sq km), forest (1,063 sq km), mangroves (2,241 sq km), marsh and tidal flats (4,750 sq km), other swamp and forested wetland (1,607 sq km), pasture (18 sq km), sandy beach (75 sq km), scrub, grasslands, prairie, sandhill (203 sq km).

CLIMATE CHANGE ADAPTATION PLANS FOR FLORIDA

Borisova et al. (2008) analyzed two studies on the effect of climate change on the economy of Florida: Stanton and Ackerman (2007) and Harrington and Walton (2007). These two studies adopted different approaches and methodologies but they unanimously concluded that the economic impacts of climate change would be significant. Stanton and Ackerman (2007) reported that the annual losses from climate change impacts due to tourism reductions, hurricane damages, real estate losses, and increased costs of electricity generation may reach $345 billion by the end of the twenty-first century. Harrington and Walton (2007) found significant loss of property value as a result of possible inundation due to sea level rise and increased storm damage. The predicted value of land at risk for 2080 could be really high in three coastal counties under study (Dade, Duval, and Escambia) with more than $10 billion in Dade County alone.

There are crucial areas where adaptation planning and implementation are needed in order to avoid, minimize and mitigate the anticipated effects of climate change (SWFRPC, 2011). In a regional vulnerability assessment, 246 climate change management adaptations were identified (Beever III et al., 2009) that could be utilized to address the various vulnerabilities identified for the region. Based upon a variety of effects analyses, the prioritization ranking for the climate change vulnerabilities is in descending order: altered hydrology, climate instability/storm severity, habitat and species changes, geomorphic (landform) changes, sea level rise and water temperature and chemistry changes, infrastructure impacts and land use changes, air temperature and chemistry changes and human health, human economy, and variable risk (Beever III et al., 2010). It is crucial to plan and act immediately to avoid, mitigate, minimize, and adapt to the negative effects of climate change, and to examine the possibilities of providing benefits to human and natural systems by adapting to the changing planet. CHNEP (2010) reported that in Florida, water resources are managed through several mechanisms. The five water management districts in Florida are the agencies primarily responsible for managing water for water supply, flood protection, water quality, and natural systems. Water Management Districts are required to adopt and update water supply plans and to establish minimum flows and levels (MFLs), forecast water demands by sector and identify needed new water supplies. New water supply alternatives within the existing water supply plans include alternative sources: captured surface water/storm water during high flows, reclaimed water, seawater desalinization, brackish groundwater desalinization, fresh groundwater, non-agricultural water conservation, agricultural water conservation, and aquifer storage and recovery. Bloetscher et al. (2011) reported that water managers must plan to optimize goals for water supply, flood control, and ecosystem protection.

Mulkey (2007) reported that Florida can become a leader in mitigation of GHGs through effective management of agriculture, forestry, and natural ecosystems. Mitigation in these sectors can significantly minimize projected increase in fossil-derived GHGs over this century. Such management will not be possible without comprehensive data on the carbon budgets and emissions of these systems. The state could develop the resources necessary to collect these data. Appropriate environmental safeguards are essential to insure that the methods of mitigation are consistent with the long-term health of Florida's ecosystems. Obeysekera et al. (2010) reported that investigations of future water management plans would require reliable information on climate projections (both precipitation and temperature) at regional scales.

CONCLUSION

While the extent of climate change is uncertain, we need to be prepared for the likely outcomes resulting from climate change by taking sensible actions, and adopting suitable adaptation strategies. Berry et al. (2011) reported that groundwater resources are not sufficient now to meet future water demand in many parts of Florida, and climate change adds to the vulnerability of groundwater supply and demand. Therefore, developing alternative water supplies is a priority. While a few municipalities and counties are planning to develop new surface water sources, the difficulties of storing water in regions with little topographic relief, porous geology, and shallow water tables limits the viability of developing significant surface water sources. Because of these limitations on traditional surface and groundwater sources, alternative sources, such as desalination and reuse, will be significant components of Florida's future water supplies (Berry et al., 2011). Conservation efforts and appropriate water management approaches are important for long-term sustainability of the valuable water resources in the

Florida watersheds. An integrated cross-agency approach should be adopted by exploring common planning factors (sea level rise, precipitation, etc.), resolving which models are best suited for the Everglades watershed, and reducing uncertainty through regional scale application of larger scale climate change models (SFERTF, 2010).

REFERENCES

Armentano, T. V., Park, R. A. & Cloonan, C. L. (1988). Impacts on coastal wetlands throughout the United States. In J. G. Titus (Ed.) *Greenhouse effect, sea level rise and coastal wetlands.* EPA-230-05-86- 0.13. U. S. Environmental Protection Agency, Washington, USA.

Bamber, J. L., Riva, R. E. M., Vermeersen, B. M. A. & LeBrocq, A. M. (2009). Reassessment of the potential sea-level rise from a collapse of the west Antarctic ice sheet. *Science* 324, 901-903.

Barnett, C. (2000*). Swamp repair: Florida's latest weapon in the battle to preserve its dwindling wetlands isn't living up to its promise,* Florida Trend; May.

Barnett, C. (2007*). Mirage: Florida and the vanishing water of the Eastern U. S.* The University of Michigan Press, Ann Arbor, MI.

Beever, J. W. III, Gray, W., Trescott, D., Cobb, D., Utley, J. & Beever, L. B. (2009). *City of Punta Gorda climate change adaptation plan.* SWFRPC/CHNEP, 267 pp.

Beever, J. W. III, Gray, W., Trescott, D., Cobb, D., Utley, J. & Hutchinson, D. (2010). *Lee County climate change vulnerability assessment.* Southwest Florida Regional Planning Council, Florida.

Berry, L., Bloetscher, F., Hernández Hammer, N., Koch-Rose, M., Mitsova-Boneva, D., Restrepo, J., Root, T. & Teegavarapu, R. (2011*). Florida Water Management and Adaptation in the Face of Climate Change*, Florida Climate Change Task Force, Florida

Bloetscher, F. (2009). Climate Change Impacts on Florida (with a specific look at groundwater impacts). *Florida Water Resources Journal.* September 14-26.

Bloetscher, F., Heimlich, B. & Meeroff, D. E. (2011). Development of an adaptation toolbox to protect southeast Florida water supplies from climate change. *Environ. Rev.* 19, 397-417.

Borisova, T., Breuer, N., & Carriker, R. (2008*). Economic Impacts of Climate Change on Florida: Estimates from Two Studies.* Food and Resource

Economics Department, Florida Cooperative Extension Service, Institute of Food and Agricultural Sciences, University of Florida.

Bradley, R. S., Diaz, H. F., Eischeid, J. K., Jones, P. M., Kelly, P. M. & Goodess, C. M. (1987). Precipitation fluctuations over Northern Hemisphere land areas since the mid-19th century. *Science*, 237, 171-175.

Cazenave, A., Lombard, A. & Llovel, W. (2008). Present-day sea level rise: A synthesis. *CR Geosci*, 340(11), 761-770.

Charlotte Harbor National Estuary Program (CHNEP). (2010). Charlotte harbor regional climate change vulnerability assessment. Florida. Available at: *http://www.chnep.org/projects/climate/Vulnerability Assessment2-19-10.pdf*

Clark, P. U., Huybers P. (2009) Interglacial and future sea level. *Nature,* 462, 856-857.

Crisci, A., Gozzini, B., Meneguzzo F., Pagliara S. & Maracchi G. (2002). Extreme rainfall in a changing climate: Regional analysis and hydrological implications in Tuscany; *Hydrol. Process,* 16, 1261–1274.

Dawson, B., & Spannagle, M. (2009). *The complete guide to climate change.* Routledge, Oxon, UK.

Department of Environmental Protection (2011). Florida's Watershed Management Approach. Available at: *http://www.dep.state.*

Diaz, H.F., Bradley, R.S. & Eischeid, J.K. (1989). Precipitation fluctuation over global land areas since the late 1800s. *J. Geophys. Res.,* 94, 1195-1240.

Erwin, K.L. (2009). Wetlands and global climate change: the role of wetland restoration in a changing world. *Wetlands Ecol Manage.,* 17, 71-84.

Ewel, K. C. & Smith, J. E. (1992). Evapotranspiration from Florida pond cypress swamps. *Wat. Resour.*, 28, 299-304.

Fiedler, J., Mays, F. & Siry, J. (eds.) (2001*). Feeling the heat in Florida: Glogal warming on the local level.* Natural Resources Defense Council and Florida Climate Alliance, New York and Orlando.

Friedli, H., Lotscher, H., Oeschger, H., Siegenthaler, U. & Stauffer, B. (1986). Ice core record of the C-13/C-12 ratio of atmospheric CO2 in the past 2 centuries. *Nature*, 324, 237–238.

Fu¨ssel, H. M. (2007). Vulnerability: A generally applicable conceptual framework for climate change research. *Global Environmental Change*, 17, 155–167.

Fitterman, D. V., Deszcz-Pan, M. & Stoddard, C. E. (1999). Results of time-domain electromagnetic soundings in Everglades National Park, Florida (on CD-ROM). *U.S. Geological Survey Open-File Report*, 99–426.

Gaffen, D. J. & Ross, R. J. (1999). Climatology and Trends of U.S. Surface Humidity and Temperature. *J. Climate*, 12, 811–828.

Gornitz, V. & Lebedeff S. (1987). Global sea level changes during the past century. In Nummedal, D., Pilkey, O. H., & Howard, J. D. (Eds.), Sea-level fluctuation and coastal evolution, *SEPM Special Publication* No. 41.

Groisman, P.Y. & Easterling, D. R. (1994). Precipitation changes over the northern hemispheric extratropics during the last hundred years. In Desbois, M. & Désalmand, F. (Eds.). *Global precipitations and climate change,* NATOASI Series, Vol. 26, Springer-Verlag Berlin Heidelberg.

Harrington J., Walton T. L. (2007). *Climate change in coastal areas in Florida:* Sea level rise estimation and economic analysis to year 2080. Florida State University, Tallahassee, Florida.

Intergovernmental Panel on Climate Change. (2007). *Climate change 2007: The physical science basis.* In Soloman, S., Qin, D., Manning, M., Chen, Z, Marquis, M, Averyt, K. B., Tignor M., & Miller, H. L. (Eds.). Contribution of Working Group I to the Fourth Assessment Report of the Intergovernmental Panel on Climate Change. Cambridge University Press: New York.

Maul, G. A. (1993). Sea level rise at Key West, Florida, 1846-1992: America's longest instrument record? *Geophys. Res. Lett.* 20 (18), 1955-1958.

Martin, J., Fackler, P. L., Nichols, J. D., Lubow, B. C., Eaton, M. J., Runge, M. C., Stith, B. M. & Langtimm C. A. (2011). Structured decision making as a proactive approach to dealing with sea level rise in Florida. *Climatic Change*, 107, 185–202.

Mearns, L.O., Schneider, S. H., Thompson, S. L. & McDaniel, L. R. (1990). Analysis of climate variability in general circulation models: comparison with observations and changes in variability in 2xCO2. *J. Geophys. Res.* 95, 20469-20490.

Mulholland, P. J., Best, G. R., Coutant, C. C., Hornberger, G. M., Meyer, J. L., Robinson, P. J., Stenberg, J. R., Turner, R. E., Vera-Herrer, F. & Wetzel, R. G. (1997). Effects of climate change on freshwater ecosystems of the south-eastern United States and the gulf coast of Mexico. *Hydrol. Process.* 11, 949-970.

Mulkey, S. (2007). *Climate change and land use in Florida: Interdependencies and opportunities.* A report prepared for the Century Commission for a Sustainable Florida.

National Research Council (2010). *Advancing the science of climate change.* Washington, DC: National Academies Press.

Noda, A. & Tokioka, T. (1989). The effect of doubling the CO_2 concentration on convective and non-convective precipitation in a general circulation model coupled with a simple mixed layer ocean model. *J. Meteorol. Soc.* 67, 1057-1069.

Neftel, A., Oeschger, H., Schwander, J., Stauffer, B., & Zumbrunn, R. (1982). Ice core sample measurements give atmospheric CO2 content during the past 40,000 yr. *Nature*, 295, 220–223.

Obeysekera J., Irizarry M., Park J., Barnes J., & Dessalegne T. (2010). Climate change and its implications for water resources management in South Florida, *Stoch. Environ. Res.* Risk Assess.

Pearlstine, L. G., Pearlstine, E. V., Aumen, N. G. (2010). A review of the ecological consequences and management implications of climate change for the Everglades. *J. N. Am. Benthol. Soc.* 29, 1510-1526.

Price, R. M., Swart, P. K., Fourqurean, J. W. (2006). Terrestrial brackish groundwater discharge—a significant source of phosphorus for the oligotrophic wetlands of the Everglades. *Hydrobiol.* 569, 23-36.

Rahmstorf, S. (2007), A semi-empirical approach to projecting future sea-level rise. *Science* 315, 368–370.

Rignot, E., Velicogna, I., Van den Broeke, M. R., Monaghan, A., & Lenaerts J. (2011). Acceleration of the contribution of the Greenland and Antarctic ice sheets to sea level rise. *Geophys. Res. Lett.* 38, L05503.

Scheraga, J. D. & Grambsch, A. E. (1998). Risks, opportunities, and adaptation to climate change. *Climate Res.* 10, 85–95.

South Florida Ecosystem Restoration Task Force (SFERTF). (2010). *Climate Change in South Florida.* Miami.

SFWMD (2009). *Climate change and water management in South Florida. South* Florida Water Management District. West Palm Beach, Florida.

Smith, J. B. & Tirpak, D. (1989) *The Potential Effects of Global Climate Change on the United States,* Office of Policy, Planning and Evaluation, U.S. Environ. Protect. Agency, Washington, D.C., EPA, 230-05-89-050.

Solomon S., Qin, D., Manning, M., Chen, Z., Marquis, M, Averyt, K. B., Tigora, M., & Miller, H. L. (2007). *Climate Change* 2007*: The Physical Science Basis. Contribution of Working Group I to the Fourth Assessment Report of the Intergovernmental Panel on Climate Change,* Cambridge University Press, Cambridge, UK.

Solomon, S., Plattner, G. K., Knutti, R. & Friedlingstein, P. (2009). Irreversible climate change due to carbon dioxide emissions. *Proc. Natl Acad. Sci. USA*. DOI: 10.1073/ pnas.0812721106.

Soule, P. T. (2005). A comparison of 30-yr climatic temperature normals for the southeastern United States. *Southeastern Geographer* 45, 16-24.

Stanton, E. A., & Ackerman, F. (2007) *Florida and climate change: the costs of inaction.* Tufts University Global Development and Environment Institute, Medford, MA (USA).

SWFRPC (South West Florida Regional Planning Council) (2011). CHNEP/SWFRPC Climate Ready Programs. Florida. Available at: *http://www.swfrpc.org/content/Natural_Resources/Climate_Change/Climate_Change.pdf*

Titus, J. G. (1988). *Greenhouse effect, sea level rise and coastal wetlands.* U.S. Environmental Protection Agency. EPA-230-05-86-013.

Karl, T., R., Melillo, J. M., & Peterson T. C. (2009) Global Climate Change Impacts in the United States. *U.S. Global Change Research Program.* Cambridge University Press, NY.

USEPA (2009). The effect of climate change on water resources and programs. United States Environmental Protection Agency. Available at: *http://www.epa.gov/owow_keep/watershed.*

Vinnikov, K. Y., Groisman, P. Y. & Lugina, K. M. (1990). Empirical data on contemporary global climate changes (temperature and precipitation). *J. Climate* 3, 662-677.

Webster, P. J., Holland, G. J., Curry, J. A., & Chang, H. R. (2005) Changes in tropical cyclone number, duration, and intensity in a warming environment. *Science* 309, 1844-1846.

Winsberg, M. D. (2003). *The controls of weather and climate and their application to Florida.* In: Florida Weather. 2nd Ed. University Press of Florida, Gainesville, FL, 40-45.

World Wildlife Fund (WWF) (2011). Florida: At the forefront of climate change. Available at: *http://test.worldwildlife.org/climate/florida/index.html.*

Zhang, K. (2011). Analysis of non-linear inundation from sea-level rise using LIDAR data: a case study for South Florida. *Climatic Change* 106, 537-565.

In: Watersheds
Editors: G. L. Kuhn and J. R. Emery
ISBN 978-1-62081-057-6

Chapter 5

FACTORS AFFECTING FARMERS' ADOPTION OF BEST MANAGEMENT PRACTICES IN ONTARIO'S GRAND RIVER WATERSHED*

Glen Filson, Delia Bucknell and Stewart Hilts
University of Guelph, Guelph, ON, Canada

INTRODUCTION

To see how recent regulations and programs have operated in the Grand River watershed, we talked to Conservation Authority officials and consulted recent water quality studies to identify the sub-watersheds with the worst and best water quality. This chapter describes how residents within the sub-watersheds of the Grand, the Eramosa/Speed (E/S) and the Canagagigue Creek (CC) perceive agri-environmental programming and recent legislation imposed after the Walkerton tragedy. Earlier research had found fewer water quality problems in the E/S region than in CC (see GRCA, Feb. 14, 2002; Beak International *et al.*, 1999). We thought this might be due to different land use practices, degrees of agricultural intensity, the attitudes of rural people toward environmental best management practices (BMPs) and the impact those practices might have on rural people's perceived quality of life.

---

* A version of this chapter also appears in *Agriculture and Environmental Security in Southern Ontario's Watersheds*, edited by Glen Filson, published by Nova Science Publishers, Inc. It was submitted for appropriate modifications in an effort to encourage wider dissemination of research.

In keeping with the conceptual framework, we recognized that the adoption of BMPs is an environmentally responsible behaviour arising out of environmental concern. "BMPs are proven, practical and affordable approaches to conserving water, soil and other natural resources" (MVCA, 2005). A comparison of residents' views about the voluntary adoption of BMPs versus the relatively new requirement that large operation farmers implement a Nutrient Management Plan (NMP)[1], enhances our understanding about whether farmers were adopting BMPs because of their environmental concerns or primarily because it was required of them.

To obtain a clearer picture of farmers' views about regulation respondents were also asked to comment on regulatory aspects of the Nutrient Management Act (NMA) (2002) and to say what they thought about voluntary environmental cost share programs like the Environmental Farm Plan (EFP) and the Grand River Conservation Authority's Rural Water Quality Program (RWQP). The survey was conducted with both farm and non-farm residents in these regions to obtain their perceptions about the effects of their local farming systems and whether they suspected large, medium or small farm operations of being the biggest causes of pollution. We tried to determine whether we could identify different levels of adoption of BMPs as a function of their farm characteristics, the reported water quality of particular sub-watersheds and/or their demographic characteristics.

FACTORS MOTIVATING FARMERS TO INTRODUCE BMPS

As early as 1991, Duff *et al.* noted that education and financial incentives are seen by farmers to be more effective than other government policies in promoting conservation. They observed that farmers tend to prefer a voluntary approach to environmental management, despite the fact that they perceive regulatory approaches to be more effective in some circumstances. Kaiser and Shimoda (1999) also observed that people feel morally responsible more as a result of guilt feelings than for the reason that they must behave conventionally.

[1] Relaxing the earlier requirement that all farmers introduce an NMP occurred in part because of the widespread perception that environmental pollution is mainly caused by so-called 'unsustainable factory farms' but the Walkerton tragedy occurred as the result of an improperly chlorinated well on a small to medium sized beef farm which had done the due diligence of implementing an Environmental Farm Plan.

Cestti *et al.*, 2003 argued that the effectiveness of BMPs designed to improve environmental conditions is a function of local conditions including 1) topography; 2) climate; 3) cropping system; 4) maintenance; 5) proper site selection and 6) proper installation (). Thus the effectiveness of BMPs has a lot to do with site specificity. This also affects farmers' motivation to adopt because if a program is not tailored specifically enough, farmers may not be interested, and this too becomes a barrier. This is also expected because there are different types of mindscapes (Maruyama, 1985). These different types of mindscapes affect the landscapes and the ecological goods and services produced by farmers as do their farm sizes, the types of commodities they produce and their overall economic well-being.

Intention is also very important in addressing the issue of who adopts environmental best management practices. According to the Hines Model (cited in Fransson and Garling, 1999), intention is a factor most closely related to actual behaviour. Intention is related to knowledge, skill and personality. Thus pessimistic farmers will only adopt best management techniques if they have the intention but not necessarily because of regulation. Intention can be developed by interacting with farmers who had already adopted the environmental management practice. However, some farmers probably implement BMPs because they are being told to do so (Filson, 1996) or receive peer pressure from family and friends to do so (Ryan, 1995).

Farm characteristics influence the adoption of voluntary conservation practices and programs in several ways. According to Gale *et al.* (1993), the following factors can have an effect on whether farmers will adopt conservation measures. First, the larger the farm, the more likely and more willing farmers are to adopt new practices, as larger farms require more information to operate and would therefore be more exposed to environmental literature and issues (Tucker, 2002 and Serman, 1999). Second, the higher the sales and income on a farm, the more likely programs will be adopted (Luar, 1993) as larger farms generally have higher incomes, and smaller farmers may be unable to commit what could be a significant proportion of their incomes to conservation measures. Third, it is hypothesized that there is less priority toward rented land and the reception of conservation program (Gale *et al.*, 1993). Fourth, the type of farm enterprise will affect the adoption of conservation practices (Morris and Crawford, 2000). Gale *et al.* (1993), for instance, found that farmers with lower economic status had lower participation rates in the American Rural Clean Water Program (see chapter 6 below).

Also, a farmers' involvement in environmental organizations means they will be more likely to adopt an environmental program (Smithers and Furman, 2003; Luzar, 1993 and Morris and Potter, 1995) as they will likely be more exposed to the effects of environmental conservation programs. Involvement in environmental organizations also affects farmers' cognition as the result of interaction. Belonging to such organizations may affect farmers' locus of control which in turn may affect their adoption of best management practices. Farmers with an internal locus of control may be better adopters and environmental managers.

Religion can also play a role in farmers' environmental consciousness and therefore affect their likelihood of adopting conservation practices. While researching the Clean Up Rural Beaches program, Ryan (1999) found that farmers belonging to the Christian Farmers Federation of Ontario (CFFO) had lower environmental attitude scores than Ontario Federation of Agriculture (OFA) members, who in turn had lower scores than lifestyle farmers. She argued that Christian theology may play a role in farmers' environmental attitudes. She noted that "The statements about the finite resources of the earth and the relationship of humans and nature appeared to cause the most concern" (Ryan, 1999: 194). After observing that the CFFO's *Earthkeeping Ontario* newsletter and policy statements reflect respect for the environment she argued that leaders may be more progressive than their members.

Stewardship or 'earthkeeping' implies a use of creation to benefit all people and creatures. The concept is that if we care for creation it will provide us with all we need to survive. This differs considerably with standard environmental ideology that tends to emphasize preservation rather than use (Ryan, 1999: 194).

Stern and Oskamp (1987) also assume that activated norms are sometimes accepted in more specific form as norms for one's behaviour. A violation of these personal norms involves feelings of guilt whereas compliance invokes feelings of pride. Religion is a very important determinant of the values and norms that prevail in any society.

Former CFFO strategic policy advisor van Donkersgoed has called for a suspension of the NMA and other regulations affecting Ontario farming (2005). While the CFFO has agreed that large farms ought to "document their stewardship" (van Donkersgoed, 2004) the CFFO feels that the rest of society should pay farmers for producing "our clean water, fresh air, biodiversity, abundant wildlife and attractive landscapes" in the form of "annual whole farm payments" (van Donkersgoed, 2005) which is what European Union farmers return for producing Environmental Goods and Services (EGSs) (see chapter 2 and 3).

Serman's 1999 analysis of farmers' perceptions of Ontario's 1980s and early 1990s Soil and Water Environmental Enhancement Program (SWEEP) found a correlation between program uptake and farm size, gross farm income and farmers' educational levels. Several other Ontario studies have been done to try to understand farmers' motivation to adopt BMPs as a function of farm and demographic characteristics (e.g. Stonehouse, 1996; and McCallum, 2003). Economic constraints, mistrust of government, insufficient promotion, a shortage of skilled agricultural extension workers and the perception that the environment is fine as it is were among the most commonly cited reasons why some farmers do not participate in conservation programs.

In 2001, southern Ontario's Maitland Valley Conservation Authority (MVCA) commissioned a study of landowner motivations particularly among its concentrated livestock producers and found relatively positive attitudes among farmers towards the introduction of conservation practices with the vast majority of interviewees agreeing with the need to fence cattle off from streams and prepare Nutrient Management Plans (Kayak, 2001). The survey found 78% of farmers willing to adopt environmental programming if there were financial incentives like cost-sharing and tax reductions with technical advice and resources to facilitate program implementation (Kayak, 2001).

Blackie and Tuininga (2003) investigated complaints in Ontario's ten most threatened watersheds resulting from manure spills recorded by the Ministry of the Environment's Spills Action Centre. The top three watersheds impacted by livestock spills and discharges were the Ausable, Grand and Maitland Rivers. Ausable Bayfield watershed had the most manure spills in Ontario with 29% of its existing manure storages undersized. Such manure management problems as winter manure spreading and failing to check tile drainage for manure discharge were common. The larger, commercial livestock producers generally had the most appropriate manure handling facilities whereas substantial underinvestment in such facilities was the norm for most of the smaller operations. Obviously there is a need to improve on the use of BMPs to not only reduce excess nutrients but to minimize pesticide use, prevent soil erosion, support biodiverse environments and limit the production of greenhouse gases such as methane produced by livestock.

One of the GRCA's most important innovations has been the development of its Rural Water Quality Program (RWQP) (reviewed in chapter 3). This cost-share program funded mainly from local municipalities with help from higher levels of government, enables farmers to qualify for funds to improve their manure storage, fence cattle from streams and introduce other BMPs like riparian buffer strips. Farmers can qualify for grants for manure, fertilizer and fuel storages, help with the development of nutrient management plans and the retirement of fragile lands. Grants can be between 50 to 100% of the cost, depending on the kind of project funded (Schultz *et al.* 2004). Since inception in 1998 over \$16 million had been spent by the RWQP and more than \$8 million by farmers on such projects (Ryan, 2006) and the amounts continue to climb.

The CC has been selected as one of five surface water priority areas in the Waterloo Region where farmers can qualify for RWQ grants (Gray *et al.*, 2004). *The Canagagigue Creek Watershed Report* (2002) discussed several issues connected with this study including erosion of watercourses through animal access, lack of vegetation along the sides of these watercourses (vegetative buffer strips), and surface run-off. This has contributed significantly to the Creek's poor water quality. Programs such as the Rural Water Quality Program, developed by the GRCA's Ryan with other rural stakeholder groups, have been successful to a degree in mitigating some potential for risk, though problems still exist.

The GRCA's Ryan (1999) has argued that there are both commercial BMPs, like manure storages which will eventually make farmers money, and purely environmental BMPs like fencing which come at a cost but yield no financial benefits. Farmers have more incentive to introduce the former but the RWQP helps farmers introduce both types as long as they have first completed an EFP. While Diffusion of Innovation theory helps explain why the commercial BMPs are introduced by innovators and early adopters, who usually operate bigger farms, have higher incomes and educational levels and are less bound by tradition, as predicted by Rogers (1995), Ryan calls for a Theory of Altruism to explain why some farmers introduce net cost BMPs in a manner often inconsistent with the characteristics that Rogers argued usually predict innovators and early adopters.

Trust between farmers and the environmental agent, whether it is with the government, an environmental organization or non-farm rural resident, is not easily earned. It is important for farmers to see that agents are willing to make significant commitments, whether it be funding or program longevity for example, as it is unrealistic to expect farmers to do the same if this commitment is not reciprocated. If a climate of reciprocity is not obtained in the form of commitments, farmers will be unlikely to trust the program that is being proposed or implemented (Marshall, 2004), and will therefore be unlikely to adopt a best management or other conservation practice.

Rogers (1995) found that information obtained from peers carries more weight than information from the media, scientists, and likely the government. Yet it is important that farmers feel they can trust the government, which will increase compliance with any programs implemented (Marshall, 2003). Ensuring that farmers are consulted, and that enough funding is available to ensure compliance will provide the building blocks of trust between farmers and the government or other agent to foster healthy stewardship practices.

The attitudes of farmers and non-farmers in a relatively good water quality sub-watershed of the Grand River (Eramosa/Speed) were compared with the attitudes of farmers and non-farmers in a sub-watershed which had very poor water quality (Canagagigue Creek) to see if there were systematic differences.

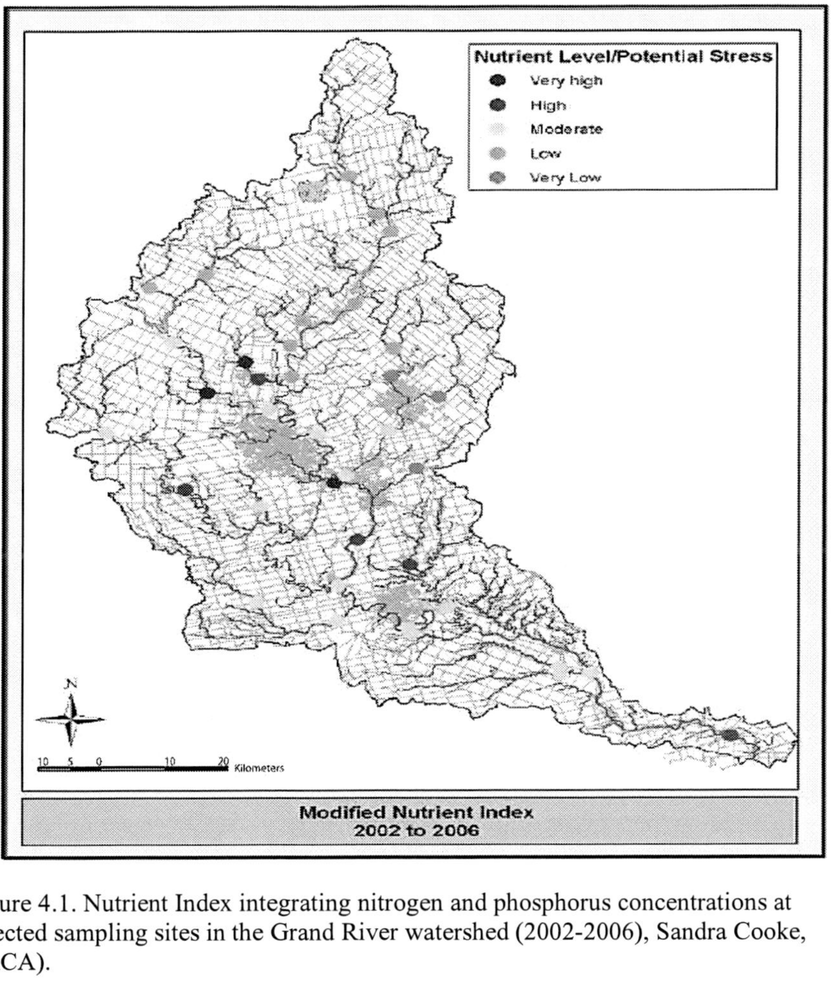

Figure 4.1. Nutrient Index integrating nitrogen and phosphorus concentrations at selected sampling sites in the Grand River watershed (2002-2006), Sandra Cooke, GRCA).

THE STUDY AREAS

Located on the outer edge of the Greater Toronto Area (GTA), the Grand River watershed has its own urban pressures from the cities of Guelph, Kitchener-Waterloo, Cambridge and Brantford.

About 800,000 people live in the watershed in part because of its prime agricultural land, favourable climate and access to markets (Schultz *et al.*, 2004). In the figure above, the E/S sub-watershed, with good water quality, passes south into Guelph whereas the CC, with poor water quality, passes from Elmira south into Kitchener/Waterloo. As the Speed River passes through Guelph, inadequate storm water treatment causes a decline in water quality but CC water quality is generally much worse than that of the E/S. This motivated us to compare the views of rural people in each area to environmental issues, BMPs and the impact of intensive agriculture on people's quality of life.

By contrast with the E/S, the area surrounding the CC is primarily rural, with a significant number of farms. The average age of farmers interviewed in the Canagagigue area was 49.3 years, with all of the respondents being male. This is similar to the average age of the farming population as a whole in the area though the average age of farmers throughout Ontario is even higher. The average farm size in the study was 117 acres and ranged from 31 to 270 acres with a median of 100 acres, which is smaller than the average of approximately 167 acres in C.C. overall. With respect to level of education, 72% of the farmers had completed grade eight, 14% had completed some of high school, and 3% had attained a diploma in agriculture. The average educational attainment in the area as a whole was higher with 30.1 % of the farmers having a grade 8 education and 26.1% having completed high school.

Farm operations that had beef cattle as a commodity were the most prevalent, with combinations with either pork or dairy being common. Poultry was the least common of all livestock operations, with only three of the farms surveyed having any poultry, whether stand alone, or in combination with other animals.

Within the Creek several factors damage water quality. Erosion is evident in 62% of CC's stream banks exacerbated by increased pasturing, tillage and cultivation along watercourses. Eutrophication of reservoirs is of concern as well, because as phosphate levels continue to increase, levels of dissolved oxygen decrease, in turn hurting oxygen-reliant organisms (GRCA, 2002).

Both point and non-point source pollution in the Creek have caused environmental problems. In 1989 it was discovered that a shallow aquifer near Elmira had been contaminated with nitrosodimethylamine and cholobenzene, as a result of leaching wastes from Uniroyal, a major chemical manufacturing company (Guelph Water Management Group, 2005). The result has been indefinite closure of many source wells and aquifers (Schultz *et al.,* 2004). Instead of cleaning up their DDT and dioxin laden-site, the company had been fighting efforts to force a clean-up to the detriment of the Mennonite farmers surrounding the plant. "Mennonites tend not to engage politically in land use and other decision-making and consequently suffer for their beliefs" (Dempster *et al.,* 2004: 157) because they often chose not to demand that Uniroyal clean up their pollution.

The GRCA's Water Resources Manager, Lorrie Minshall also indicated that more than half of the water quality problems in the CC are due to non-point source pollution, largely from agriculture. Run-off from fields, barnyards and improper manure handling and storage are among some of the issues that must be addressed. Other issues include animal access to stream banks and loss of topsoil, representing significant non-point sources of pollution (Minshall, Oct. 3, 2002). In 1999, the CC was identified as a high priority for the reduction of agricultural pollutants. The following livestock farms were considered high priorities for the RWQP (table 4.1).

Comparisons of Eramosa (within E/S) and Peel Townships provide useful insight to the differences between E/S and CC. Peel has been amalgamated with Maryborough and Drayton Village to form Mapleton Township. Generally the landscape in the Eramosa region is more rolling, with drumlins and wetlands in lower valleys, and a higher proportion of remaining forest cover. In the former Peel Township, in the upper CC watershed, the landscape is much more flat, with a very low proportion of remaining forest cover.

Table 4.1. Non-Point Course Inventory of Canagagigue Creek Problems

Total Livestock Farms: 200 dairy, 78 beef, 25 swine, 19 poultry, 8 horses	315
High priority manure storage needed	109
Medium priority manure storage needed	151
Low priority manure storage needed	55
Projected milkhouse impacts	66 (1/3 of dairy farms)
Livestock access sites to the Creek	140
Tillage in floodplain	13
Sites with visible soil erosion	293
Sites requiring buffers	272

(Source: GRCA, Feb. 14, 2002).

CC farms are considerably more intensive, in terms of both crops and livestock, than in the E/S region. In a study of the biodiversity within the region, Okey (1998) found a relatively high density of livestock in the CC sub-watershed compared with other parts of the Grand River watershed.

According to Statistics Canada (2001), in Mapleton there are 491 farms that carry cattle and calves; 336 farms in Woolwich and 221 farms in Centre Wellington that carry cattle and claves. Pigs are also raised a lot in Mapleton with 159 farms; 138 farms in Woolwich and 69 farms in Centre Wellington. There are also 40 farms in Centre Wellington that raise sheep and lamb; 36 farms in Mapleton and 26 farms in Woolwich. Other livestock include variables such as horses and ponies-where Woolwich has 223 farms that raise horses and ponies; Mapleton has 187 farms and Centre Wellington has 92 farms. Lastly, the poultry inventory includes the number of farms that raise chickens. Mapleton has the greatest number of farms that raise hens and chickens with 227 farms; Woolwich has 207 farms and Centre Wellington has 74 farms that have chickens. Overall, Mapleton seems to concentrate largely on livestock and poultry compared to other areas (Statistics Canada, 2001).

Peel Township (now Mapleton) has 83% of its farmland in crops while Eramosa Township has only 62.5%. Conversely, Peel has only 7.9% of farmland in 'Other' (mostly forest cover), the lowest in the county, while Eramosa has 22.3%, the highest. Peel Township also has intensive livestock farming, with cattle averaging one for every 2.5 acres, and pigs one for every 0.87 acres, while in Eramosa these averages are one per 4.0 acres, and one per 2 acres. This results in an even larger difference if manure production from this livestock is considered. Whether measured in terms of biological oxygen demand, nitrogen loading, or phosphorus loading, Peel Township exceeds Eramosa Township by a ratio of 5:1.

These fundamental differences in the agricultural landscape underlie the differences in farming practices and in attitudes found in this study. The CC is simply far more intensive agriculturally, despite their relatively small average farm size, than E/S especially with respect to livestock.

By contrast, Beak *et al.*, 1999 reported that the Eramosa River and its tributaries have high water quality with the Speed River having only slightly worse quality. About 45% of the land is agricultural. The improvement of the environmental quality of the area was attributed to the conservation ethic of landowners and their support for programs conducted in the region and the unique physiography, hydrogeology and vegetation (Bucknell, 2002).

Of the 11,065 people within the Guelph/Eramosa Township 90% are English only, 1% French only and over 8% many other language groups.

Thirteen percent were foreign born. About 50% are Protestants, 30% Catholic and remaining 20% are either not religious or are 'other Christians,' Jews, Sikhs or Buddhists (Statistics Canada, 2001).

CC lies mainly within Woolwich in Waterloo Region, Mapleton and Centre Wellington Townships (formerly Peel and Pilkington). Of the 27,060, including those who live in these townships and towns, about 75% were English only, 28% had a language other than English or French. German Canadians are next followed by the Dutch. Seventy percent of these regions are Protestant, 16% Catholic and 14% belong to some other religion or none at all. A sizeable percentage of the Protestants whose first language is not English are Mennonite (Statistics Canada, 2001).

The C.C. region is unique in that much of the land is dedicated to agriculture and that a significant number of Mennonites reside in the area. In the CC area there are 550 Mennonites in Centre Wellington; 2,990 Mennonites in Mapleton and 4,270 Mennonites in Woolwich (Statistics Canada, Sept. 2001).

There are at least 18 separate groups of Mennonites within Ontario. Approximately 66% are within Waterloo County where Canagagigue Creek exists (Wandel, 1995). From the early 1800s Old Order Mennonites (OOM) have been settling in what is now Mapleton, Wellesley or Center Wellington Township. Between 25 and 35% of the population in Waterloo County are Mennonites (Wandel, 1995), while the Mennonite population within Mapleton and Woolwich Townships specifically, is approximately 24% (Filson, 2004). The area includes conservative Old Order Mennonites (David Martin, Elam Martin and Old Colony), moderate Mennonites (for example, Evangelical and Beachy Amish) as well as progressive Mennonites (e.g. Mennonite Conference) (Fretz, 1989).

Typically they are committed to a 'plain' life of hard work, thrift, pacificism and a reluctance to be involved with government (and hence government programs like the RWQP). The diversity of groups in the Mennonite faith vary from the Old Order Mennonites (OOMs), where no electricity, no automobiles and strict separation from the outside world is enforced, and so-called Black Bumper Mennonites, who may drive a black car to church, though still remain relatively traditional, to the other end of the spectrum where Mennonites are indistinguishable from other farmers or citizens. For this latter reason, it has become essential for GRCA conservation stewardship extensionists to carefully explain the functions of the program to the Mennonite religious leaders because in the first years of the RWQP there was very little uptake due to the GRCA's association with municipal government. About 75% of the RWQP have been done by the Dave Martins group whereas some OOMs, who will do the EFPs (farmer based), will not take the RWQP money because of the government funding which is attached (interview with Tracey Ryan, Oct. 3, 2002).

The core area of Old Order Mennonites and other Mennonite groups live in the western part of the Grand (see the map above) where they constitute between 25 and 35% of the population (Wandel, 1995). Thus a strong OOM culture (Arthur, *et al.*, 1998) continues in the CC area though there are many non-Mennonites as well. Eramosa, on the other hand, has a broader religious and cultural base (Statistics Canada, 2002). The presence of the Old Order Mennonite group in the CC area was a factor in the research and was considered an important socio-cultural factor helping explain the findings (see the second phase of this research below).

RESEARCH METHODS

The data for this study were collected initially by conducting two focus groups, one with farmers, the other with non-farmers in the E/S region. The classification of land in the Municipal Assessment Roll was used to indicate whether or not a person was a farmer. Questionnaires were then completed via mail survey in 2002 by 385 randomly sampled farmers and non-farmers within the E/S and CC sub-watersheds of the Grand River. The questionnaire was based on the issues raised in the focus groups, the extent of use of BMPs, environmental attitudes and the impact of farmers' practices on perceived quality of life. Frequency distributions of the responses to questions were calculated along with inferential non-parametric statistics.

Relatively lower uptake of environmental BMPs, combined with our knowledge that CC sub-watershed is the Grand's more severely degraded sub-watershed, led us to conduct phase two: 29 further face-to-face, telephone interviews and an additional survey in that region. Most of these respondents wore black clothing but the questions were not about religious issues but instead, their degree of participation in EFPs, BMPs and nutrient management.

The sample in the E/S watershed was collected from the Municipal Assessment Roll at the Eramosa Township office using a random selection process. Individuals were included in the sample if they fit the criterion of living within a rural area in the region. Six hundred surveys were sent out in the E/S region, 300 to farmers and 300 to non-farmers, in 2002 but some were returned and others felt they did not qualify, leaving the total sample at 585.

Thirty-two percent responded including 66 farmers and 120 non-farmers. This under-represents female farmers and may over-represent those most adept at completing questionnaires who have with relatively higher educational and income levels.

The sample consisted of 53 women and 130 men including 9 farming females, 44 non-farming females, 55 farming males and 75 non-farming males. Most respondents in this region had completed high school and there was no significant difference in the education of the farming and non-farming groups. The mean age of all the respondents was 55.9 years with a range from 26 years to 91, which is significantly older than in CC but more typical of Ontario's farmers as a whole. Most respondents had gross incomes of less than $39,999.

A total of 651 surveys were sent out in the CC sub-watershed (minus the town of Elmira); 18 of these were returned or were received by people who did not feel they were qualified to respond leaving a total of 633 with a response rate of 31.5 percent. The sample consisted of 86 (43.2%) farmers and 133 (66.8%) non-farmers. The gender division was similar to that of the E/S study with 60 females, 133 males and 6 who did not indicate their gender. The education level of respondents, both farmers and non-farmers, in CC was significantly lower than in the E/S sub-watershed. The mean age in CC was 50.6 years, and the range was larger with ages of from 12 years to 94 years. Surprisingly a significant difference in income levels was seen between farmers and non-farmers in this region with non-farmers tending to be in the lower income brackets. Though most of the farmers were in the lower income bracket, farmers had higher incomes on average than the non-farmers.

Based on the size of the sample for the total rural populations of the two areas, the response rate from the two regions for the questionnaires had sampling errors of about 7% in CC and 5% in E/S but the sampling error is higher for specific groups such as female farmers. Given the nature of the data, variable partitioning using discriminant analyses was ruled out in preference for nonparametric statistics which require the fewest assumptions about the findings. Non-parametric statistical analysis was used because they do not require perfect probability sampling techniques, normally distributed data and the use of continuous variables.

In the first phase, farmers were asked if they would like to receive a summary of the research. When they received the summary results they were asked if they would be willing to participate in face-to-face interviews regarding government regulatory policies and environmental practices on their farms in exchange for soil tests of their land. Five farmers were interviewed in person and 18 were interviewed by phone. Help from the director of the Mennonite Cultural Centre in St. Jacobs generated an additional six surveys with farmers and a detailed soil tests were completed for interested farmers.

The mean age of the smaller group selected for intensive interviewing was 49.3 and all were male. Farms among this latter group were 30% smaller (117 acres) compared to the average of 167 for C. C. as a whole (Statistics Canada: Agricultural Census 2001). Only 3% had an agricultural diploma, 11% had completed high school, 14% had some high school education, while 72% had completed grade eight only. Most farms were small mixed livestock (beef, dairy, pork and to a lesser extent poultry) and cropping operations with grains and oilseeds followed by winter wheat, fruit and vegetables.

The main objective of the second phase of the study, conducted in the winter of 2004, was to study C.C. farmers and non-farmers' attitudes and beliefs regarding environmentalism in the C.C. sub-watershed of both farmers and non farmers. It also sought to assess farmer attitudes toward environmental programs and legislation including the NMA, EFP and BMPs. We studied the effects of the new regulations on farm operations in addition to compiling a social profile of farmers and their farms in C.C. This was particularly important due to our small sample and the need to determine whether the sample was reasonably representative of farmers' demographic and agricultural operations in the area. The social profile was used to aid our understanding about the characteristics of the study population and explore cultural differences about adoption, or lack thereof.

For the follow-up interviews conducted in 2004, a sub-sample was taken from the 2002 sample. The original farmers were contacted from a reply card from the study completed by Bucknell in which they were asked whether they would be willing to be contacted to gain further information about the results of the study. Five of the farmers agreed to being interviewed at their farms in November of 2003. Other farmers were contacted by telephone from a list which was provided from the original study asking about their farming and environmental practices. A third attempt to increase the sample size was made when the Director of the Mennonite Cultural Centre in St. Jacobs handed out surveys, with six farmers returning the surveys. The total number of farmers in the sample was 29, while the number of non-farmers interviewed was eight. The latter were representatives of environmental organizations, conservation authorities and non-farm rural residents.

Along with her return of the uncompleted questionnaire, one OFA member asked "Do Mennonite people of any sect take part in such surveys?

No!" She then went on to provide a nice overview of the issues discussed in the questionnaire and concluded by saying:

If there is a place for 'ecological goods and services' wherein the public can partake, without adding to the tax bill, without benefiting those who take without giving in society and whose mandate is clearly stated, understood by all, then we would probably support it. We do believe in having our land remain productive, with little or no erosion, in having clean water. We use a cistern and roof(s) rainwater for almost everything but drinking and will continue to do our best to follow anti-pollution rules and regulations.

The findings from the in-depth interviews, conducted to learn more about the specific issues for this group with the high percentage of Old Order Mennonites are summarized below (see Wells, 2004).

More Findings

The study assessed whether respondents in both sub-watersheds felt that their perceived quality of life was influenced by the environment as well as agricultural policies such as the Nutrient Management Act (2002). Non-farmers in both Eramosa/Speed and Canagagigue Creek regions tended to find the scenic quality of the environment to be more important than did their farming counterparts (E/S: $\rho= 0.197$, $p\leq 0.008$; CC: $\rho=0.159$, $p\leq 0.025$). Non-farmers within CC felt that environmental quality was more important to their quality of life than did farmers ($\rho=0.144$, $p\leq 0.044$). Also, non-farmers in both regions felt negatively about air pollution created by other peoples' activities. In the E/S region non-farmers were also more likely to indicate that they were somewhat negatively affected by both air pollution and by water pollution from other landowners as well as their own activities.

The Freidman test was used to ascertain whether there were any differences between farmers and non-farmers in the amount of impact on their perceived quality of life from any one type/source of pollution. The variables tested were: water pollution from others, water pollution from own activities, air pollution from others, air pollution from own activities, soil erosion/degradation, and soil compaction. In both regions farmers were more concerned about the soil and water pollution from others and non-farmers were also more concerned about the air and water pollution from others.

Respondents were asked to indicate their level of agreement on a five point scale from strongly agree to strongly disagree with thirty-one statements dealing with agricultural and environmental issues. All of these variables were tested using Spearman's rho correlation. More non-farmers agreed with the environmentally conscious statements than farmers and those who were younger, female and had a higher education also tended to be more environmentally conscious. This result is consistent with the findings of the review of literature conducted by Fransson and Garling (1999), although the impact of gender on environmental concern was ambiguous. Arcury and Christianson (1990) suggest that men are more concerned than women while Stern *et al.* (1995) asserts otherwise as does Filson (1996). Gender differences may indeed be site specific.

Respondents' attitude towards the size of farm operation was gauged by asking respondents to indicate what they saw as more of a threat to the environment - many smaller farms, fewer larger farms, both or neither. In the E/S region, non-farmers and farmers agreed that both small and large farms are a potential threat to the environment while in CC both farmers and non-farmers agreed that fewer larger farms represent more of a threat (see Table 4.2 below).

Thus most respondents were relatively evenly split between those who feel that fewer, larger farms are more of an environmental threat than both small and large farms but few people thought that many small farms are as environmentally problematic as fewer larger farms, or a combination, are.

Table 4.2. Size of Operation Perceived to be More of a Threat to the Environment

	Eramosa/Speed				Canagagigue Creek			
	farmer		non-farmer		farmer		non-farmer	
	Which type of operation is more of an env. threat?		Which type of operation is more of an env. threat?		Which type of operation is more of an env. threat?		Which type of operation is more of an env. threat?	
	count	%	count	%	count	%	count	%
many small farms	3	4.7	2	1.7	5	6.2	8	7.1
fewer larger farms	25	39.1	44	37.3	39	49.1	49	43.8
both problem-atic	26	40.6	58	49.2	31	38.3	45	40.2
Neither	10	15.6	14	11.9	6	7.4	10	8.9

Similar results were found in both study regions on the topic of government regulation of agriculture. The E/S data suggested that non-farmers in that region typically favour government regulation of agriculture to protect the environment yet farmers would prefer not to have more government regulation of agriculture.

The same tendency was found in CC with one difference; non-farmers tended to agree with farmers that 'government regulation is (not) the only way to ensure it works' (to protect the environment).

Farmers from both regions felt that government intervention in agriculture would affect their quality of life negatively; however, a majority of farmers still indicated that regulation regarding water quality could have a positive impact. The majority of farmers also felt that increasing the educational qualifications of farmers would protect the environment and that the added cost in both time and money was worth it, although they tended to disagree or were unsure as to whether government regulation was the only way to achieve environmental protection. Education is important because farmers become more environmentally responsible when they know the consequences of their actions. In contrast a higher percentage of non-farmers than farmers wanted to see increased government intervention into farming practices to protect the environment.

Both farmers and non-farmers felt voluntary cost share programs implemented by farmers such as the EFP were needed to increase environmental stewardship. The Kruskal-Wallis test of independence revealed that those who had participated in environmental programs were more likely to support 'soil testing in order to manage nutrient level' than were those who had not participated, so such interaction shapes one's self and affects one's cognition. Sixty-five percent of farmers said they had participated in an environmental program, and within the sample of farmers 45.5% had completed the Environmental Farm Plan (EFP) which, at the time was much higher than the county total (27.1%), or cash crop producers for Ontario (36%) who have a higher rate of EFP completion than most livestock producers and especially small, mixed farming operations (FitzGibbon *et al.*, 2004).

The analysis of the environmental farming practices of the CC generally revealed a low uptake of BMPs, but a reasonably high degree of participation in environmental programs. Five more farming practices were adopted by those who had participated in environmental programs than by those who had not including the creation of tree windbreaks and block plantings, the employment of conservation tillage, the improvement of manure storage/ handling facilities, development of erosion control structures, and the establishment of ditch and stream bank protection. Correlations were found with all of these practices that indicated that those who had participated in environmental programs were more likely to have employed these practices. However, the participation rate in environmental programs in the CC region was significantly lower than in the E/S region (P#0.016).

FOLLOW-UP INTERVIEWS AND SURVEYS IN THE CANAGAGIGUE CREEK (CC) AREA[2]

Almost half of the farmers (14) said there were no environmental problems in the CC area worth worrying about. Another 34% (10) said there were problems they should worry about and the remaining five did not know. When asked about what other farmers in the area were doing, 23 (79%) felt that farmers in the C. C. area were doing their part to maintain environmental sustainability, 5 (17%) believed that some were not and only 1 (3.4%) thought that farmers were definitely not managing their farms in an environmentally sustainable way. Of the 20 farmers with cattle adjacent to CC, 60% (12) fenced their cattle off from the stream while 40% (8) did not. Despite pressure to fence his cattle off from the stream, one farmer said that he would "fight it every step of the way" (Wells, 2004).

While some of these farmers were adamant that their neighbours were doing what they could reasonably be expected to do to protect the soil and water, others felt that they were doing "time honoured practices that their ancestors had be doing with no apparent ill effects" and a few were very opposed to growing pressure to change their practices. Most farmers were unable to identify any significant adverse effects that could be traced to their farming operations and so they were unwilling to participate in 'unnecessary' environmental programs. This is despite the fact that phosphorous and nitrogen production from manure are especially high in the C.C. areas at anywhere from more than 2 Kg/ha to 20kg/ha (Statistics Canada, 1996).

When asked whether there were any environmental problems in the area, or if they were worried about any specific problems, half of this study's farmers felt that there were no environmental problems to be concerned about. Another 33% felt there were some environmental problems, most commonly regarding water issues. One farmer mentioned drought, and four farmers mentioned manure run-off. Seven people (24%) pointed to the former

[2] For an in-depth review of the follow up interviews, see Wells (2004).

Uniroyal Chemical's serious pollution of the C.C. and others raised such issues as urban waste and automobile pollution. These environmental problems were seen as much more problematic than farmers' nutrient management. Regarding the barriers to their participation in environmental programs 38% point to the lack of money, 24% saw no barriers, 17% felt there was no need to change, 7% said there was too much regulation and/or problems with government run programs.

Though 41% of the farmers were unaware of any governmental conservation programs, 39% of the farmers interviewed had completed at least some stage of the EFP. Fifty-nine percent of the farmers had implemented at least one BMP while the others said they had not. Forty-seven percent had implemented windbreaks, 35% had manure storages, 5 had created buffer strips, 24% had improved their soil management and also improved their pesticide storage, 18% had implemented cleaner water management, 12% had fenced their cattle from streams, conducted intercropping and conventional tillage and 6% had done wildlife management and put in fence rows.

Regarding environmentalism, having motivation to adopt means having an incentive to do so, whether it be financial (tangible) or for stewardship reasons (non-tangible). Providing enough of an incentive for a farmer to make a behavioural change can be a difficult task as we are asking them to alter what could potentially affect their livelihood. The greatest motivational factor in farmers' decisions to adopt BMPs was that of seeing the need to make changes. Despite the water's poor quality, many farmers interviewed in this study sounded indifferent to the potential negative effects of farming on the environment.

Farmers were reluctant to see the benefits of implementing BMPs because they felt they were already being good soil and water stewards. It is important to note that two of the farmers admitted that they were "laughed at" by other farmers when they decided to implement environmental practices, so peer pressure can also have the opposite effect. Money, time, making BMPs mandatory and having more energy were all important in farmers' decisions to potentially adopt BMPs. The farming workday is long, and many farmers have off-farm employment, or other on-farm responsibilities in addition to their regular chores, which can provide little motivation. Many felt that by participating in environmental programs, more time would be diverted from income earning pursuits, and may result in a net economic loss.

Money is again a central feature in farmers' opinions regarding the adoption of BMPs. Farmers felt that without significant sources of funding such as grants from outside sources, implementing BMPs is too costly.

Structures such as manure storage systems can be expensive, and without incentives available, adopting BMPs may adversely affect the farms' profitability and therefore their livelihoods. Some Mennonite farmers' unwillingness to accept grants for religious reasons is a fundamental concern, as they would be responsible for the entirety of the cost of the project, having a greater impact on the farm's profitability. Another barrier was that some farmers simply saw no need to change their farming practices (Wells, 2004).

Over half of the farmers relied on farming as their sole support system. Another 31% said that 75% to 99% of their income comes from farming and another 14% indicate farming accounts for 74% or less of their income. On average, the number of farmers in the study who work off-farm, or have other employment, was similar to those in the C.C. as a whole. Both men and women have increased their rate of working off the farm since 1990, and roughly equal proportions of women farm operators (45.6%) and men (44.2%) worked at non-farm jobs in 2000 (Statistics Canada, 2001). This factor within the study population demonstrates the need for a financial support for especially the least well off farmers to adopt BMPs.

The amount of work associated with the implementation of BMPs was also perceived by some to be a significant barrier. The paperwork involved, permits required and the time spent building or implementing the project were all perceived barriers. Several farmers were concerned that by adopting BMPs, their yields would be adversely affected. This could occur, farmers believed, if resources were diverted toward environmental conservation and away from the operation of the farm. Or, by creating buffer strips, they would also forfeit land for cropping. This raised farmers' concerns that these conditions might adversely affect their quality of life. Farmers felt that seeing the need for change is both a barrier and a motivation for adopting BMPs.

Support for environmental programs and for further government involvement was evenly split amongst the farmers. Half of the farmers' felt that the government was sufficiently involved with farmers' environmental practices as is, and another 37% felt that the government should be helping more, mostly through financial means like availability of grants and further informational support.

The EFP had been adopted by 39% of the farmers interviewed, which was somewhat higher than the average for the province. Farmers provided varying reasons for participating in the program such as to become eligible for grant money, observing peer adoption and concern about whether their farm operation was adversely affecting the environment. Another 60% had not started the EFP either because they felt it was unnecessary or that by doing it, they might lose control of their property, or in some measure be behaving inconsistently with their faith.

The NMA was a contentious issue for only a minority of farmers in the study. Few of the farmers interviewed had implemented a NMP, and most respondents had smaller operations with fewer than the 300 nutrient units at the time they were required to complete a NMP. Two-thirds of the farmers interviewed said the NMP was a good idea, however, the remaining third felt that they would only implement it if was made mandatory by government regulation for them as well. Still others felt that it would not affect them at all, as their operations were too small to ever fall under the NMA requirements.

Discussion

The E/S region has less intensive farming including lower livestock densities so the residents experience less pressure to improve their water quality. Within CC, since Mapleton township carries a large portion of farms, this area also tends to have more livestock and Woolwich has the most crop production. Most of the rural communities are in the Mapleton and Woolwich areas where the numbers of Mennonites are the highest. The lack of information and awareness about the need to change practices, attitudes opposed to the acceptance of agri-environmental grants and a relative low level of education inhibit environmental program participation. Though again a site specific variable, religion can affect the degree to which people implement BMPs.

On the other hand, there are quite a few Old Order Mennonites who are very environmentally friendly and have become certified organic producers. However, for the most part the Mennonites participating in this study have small operations, low gross farm sales and lack the funding to implement BMPs even when so inclined.

The interviews conducted simultaneously with farmers within the CC sub-watershed of the Grand River revealed a much lower average level of education than E/S and a substantial lack of awareness about cost-share programs and even the EFP. This is unfortunate as a cost-benefit analysis of the CC study has shown that with even a small, uniform decline in net farm income of 3-5%, a sizeable reduction of sediment loading would occur if a zero tillage system were introduced while the net social benefits if implemented for the whole Grand River watershed could exceed $1.2 million (Filson *et al.* 2004).

The E/S area residents' high level of understanding of the need for and benefits from their environmental practices is probably a function of their much higher average educational level relative to the CC area. In CC many farmers do not feel they need to change their practices possibly because traditional practices are sacrosanct. Non-farmers on the other hand know there are problems in the CC area relating to agriculture though they, along with some of the farmers, insisted that the larger scale farms, or at least a combination of a few large scale farms and many small sized farms were usually the biggest sources of manure spills and other pollution problems, not many small sized farms, which is false (also see chapter 6). Agriculture is more intensive in CC than E/S and this is especially true of intensive livestock production so there is a greater need to adopt environmental BMPs in the CC than the E/S sub-watershed.

Over half of the farmers who understood that there are environmental problems in the area had adopted an EFP. The fact that the more environmentally conscious ones are the farmers who adopted these programs comes as no surprise. The concern here lies with those who have not implemented an EFP or other program, as they simply do not see any need to do so. The main advantage of the EFP relative to other plans is that it was developed by farmers for farmers. While it has been well received by most Ontario's farmers, the problem has been that many farmers have completed only the initial stages of the plan (Smithers and Furman, 2003).

Because the RWQP, which provides funding for the adoption of BMPs, is tied to the adoption of the Canada-Ontario Environmental Farm Plan, some farmers may not participate in the RWQP because they have not done an EFP. On the other hand, many have completed an EFP because they knew they would not qualify for RWQP funding otherwise.

The NMA is meant to ensure farmers are managing their nutrients effectively. While many farmers in this study were not affected when interviewed, changes in the regulation will affect farmers if they were to expand their operations. With respect to regulations farmers in the study felt overall that mandatory regulation would be the only option if stricter environmental regulations were introduced, otherwise, farmers would not adopt. Gale *et al.* (1992) found that farmers agreed that voluntary approaches alone do not work very well and that mandatory regulation may be the best way to ensure farmers adopted conservation practices, however, a balance is required between mandatory and voluntary measures.

This chapter identified factors that inhibit farmers' adoption of BMPs that were not found in studies of the other southern Ontario watersheds: the lack of education and farmers whose religion affects every aspect of life. This underlines the fact that demographic factors impacting farmers' mindscapes with respect to how they interact with their landscapes, is site specific. In some cases, religion and the lack of formal education appear to be in contradiction with the need to introduce environmentally beneficial practices.

Farmers who had a higher level of education were generally more likely to have a better awareness of government programs than other farmers. Cultural aspects may also play a role in a farmer's decision to adopt. There are different groups of Mennonites and many subscribe to very strict Old Order beliefs, even though many also are quite indistinguishable from Protestant or other farmers in their acceptance of modern technology (Stevenson, 2003). Mennonite farmers generally felt that they are good stewards of the soil whereas many non-farmer participants did not feel that most Mennonite farmers in the area were good stewards. Many Old Order Mennonites' refusal to accept government grants for fear that it may put their pacifism at risk in the event of a war is an important factor which partially explains the low adoption of BMPs and participation in government-sponsored programs.

CONCLUSION

Several contradictions were assessed in this chapter including those that exist between farmers and non-farm rural people, between traditional farm practices and contemporary requirements, between numbers of livestock and the limited amount of land to spread the excess nutrients. Farmers in both sub-watersheds of the Grand were found to have a more instrumental view of the environment than non-farmers who expressed more concerned about protecting scenic qualities and biodiversity than farmers. The differences in perception are evidence of different mindscapes associated with being a farmer versus a rural non-farmer. While farm practices generally were perceived to positively affect residents' quality of life, some occasional farm practices such as the use of sewage sludge as fertilizer are objectionable to non-farm rural residents.

A number of factors affect both what motivates farmers to adopt conservation practices such as BMPs, as well as the barriers which may prevent them from adopting. Farm or other demographic characteristics such as farm income, education and religion play a significant role in these decisions. Overall, financial constraints are the most important factors affecting farmers' decision-making, as the lack of available funding available to farmers to make the necessary changes to their operations, is admitted to or complained about by many farmers. Legislation is of concern to farmers, despite the fact many have operations that are still too small to be regulated by the Nutrient Management Act. Many farmers feel that they have to look over their shoulders to make sure Big Brother is not watching them.

Non-farmers tend to believe that existing programs such as the RWQP and the EFP are beneficial tools to help farmers to understand and implement conservation measures on their farms. They recognize, however, the importance of incentives and consideration of social characteristics in the area, and whether programs or legislation would be successful. They also note, particularly in the Canagagigue Creek which does have such poor water quality, that it is important that players from a variety of sectors participate in mitigating the effects of pollution.

Whether a farmer will adopt a conservation practice has to be considered from many perspectives. This includes whether they trust the agency delivering the program or legislation, whether voluntary or mandatory, whether they see their peers are participating or not, whether they see the rural landscape in more of a multifunctional or unifunctional sense, farmers may or may not adopt BMPs or participate in an EMS. Because agriculture in Canada is generally not seen as multifunctional as it is in Europe, the positive externalities, such as wildlife preservation and land retirement are often not considered to be a benefit, regardless of whether funding may be provided to the farmer. Changing the concept of what is a benefit, and therefore introducing the idea of a multifunctional rural landscape, is important if environmentalism is to be adopted by the more reluctant farmers.

There are relationships among the farmers' demographic characteristics, farm structural variables, awareness, motivation and the degree of environmental program participation and BMP adoption. And while there is a widespread rural perception that a few large farm operations are more of an environmental threat than many small farms, the latter often provide a chronic threat especially when they under-invest in manure management though large farm spills can be more acute when they happen (Blackie and Tuninga, 2003).

Commercial BMPs are usually initiated first by large operators with the most income, farm sales and numbers of commodities produced. Those introducing the commercially viable BMPs tend to be more aware of the impact of agriculture on the environment and on people's perceived quality of life so they are more likely to participate in environmental programs like the EFP and RWQP and have a greater interest in adopting BMPs.

More non-farmers than farmers support mandatory nutrient management and the newly established agricultural regulations but of course it is the farmers who must implement nutrient management and BMPs and to do so they need society's help. On the other hand, most farmers would prefer voluntary programming and generally resent growing government intervention in agriculture because it interferes with their independence, creates more work for them and costs them scarce money. So the conflict between farmers and governments as well as between farmers and urbanites/exurbanites contributes to a contradiction between the nature of our present food production and the environment.

REFERENCES

Arcury, T. A., and Christianson, E. H. 1990. Environmental worldview in response to environmental problems: Kentucky 1984 and 1988 compared. *Environment and Behavior*, 22(3), 378-407.

Beak International and Aquafor Feech Limited in Association with Mark Dorfman, Schroeder and Associates, Grand River Conservation Authority, Geomatics International Inc. Schollen and Company Ltd. Stantec Consulting Ltd. 1999. *Eramosa-Blue Springs Watershed Study Report.* Cambridge: Grand River Conservation Authority.

Blackie, M. and Tuininga, K. 2003. Environment Canada-Ontario Regional Compliance Promotion 2002 Target Sub-watershed Survey Report. Ottawa: Environment Canada.

Bucknell, D. 2002. *Rural Quality of Life in a Changing Environment: A Study of the Eramosa-Speed Region.* Guelph: University of Guelph MSc thesis.

Cestti, R., Srivastava, J. And Jung, S. 2003. *Agricultural Non-point Source Pollution Control.* Washington: The World Bank.

Christian Farm Federation of Ontario, 1991-2002. *Earthkeeping Ontario.* Guelph, ON.

Conservation Ontario, 2001. *The Importance of Watershed Management in Protecting Ontario's Drinking Water Supplies.* Cambridge: Grand River Conservation Authority.

Dempster, T., Sanchez-Jerez, P., Tuya, F., Fernandez-Jover, D., Bayle-Sempere, J., Boyra, A. and Haroun, R. 2004. Coastal aquaculture and conservation can work together. *Inter-Research.* 314: 309-310.

Filson, G. C. 1996. Demographic and Farm Characteristic Differences in Ontario Farmers' Views about Sustainability Policies. *Journal of Agricultural and Environmental Ethics* 9(2): 165-180.

Filson, G. C. 2004. *Annual Project Report* - 2004 of the Rural Sociological/Agricultural Extension Component of the Project: Development of Watershed System Capacity Procedure for Water Quality.

Filson, G. C., D. P. Stonehouse, R. Rudra and P. Voroney, Jan. 15. 2004. Improving sustainability and nutrient management of an agricultural watershed, *Final Report,* Guelph: Univ. of Guelph/OMAF Resources Management and the Environment Program.

FitzGibbon, J., Plummer, R. and Summers, R. The Ontario Environmental Farm Plan: A Whole-Farm System Approach. In Filson, G. C. 2004. *Intensive Agriculture and Sustainability: A Farming Systems Analysis.* Vancouver: UBC Press.

Fretz, J. W. 1989. *The Waterloo Mennonites: a Community in Paradox.* Waterloo: Wilfrid Laurier Press.

Gale J.A., Line, D. E. Osmond, D. L. Coffey, S. W., Spooner, J. and Arnold, J. A. 1993. Evaluation of Experimental Rural Clean Water Program. In Environmental Protection Agency report (section: *EPA-841-R-93-005).* Environmental Protection Agency.

Guelph Watershed Management Group, Assessing Collaborative and Integrated Water Management in the Maitland River Watershed. Waterloo: University of Waterloo.

Fransson, N. and Garling, T. 1999. Environmental Concern: Conceptual Definitions, Measurement Methods, and Research Findings. *Journal of Environmental Psychology. 19:369-382.*

Grand River Conservation Authority. 2002. *A Watershed Plan for the Canagagigue Creek Watershed, Phase I.* Cambridge: Grand River Conservation Authority.

Kaiser, F.G. and Shimoda, T. A. 1999. Responsibility as Predictor of Ecological Behaviour. *Journal of Environmental Psychology.* 19:243-253.

Luzar, E.J. and Diagne, A. 1999. Participation in the Next Generation of Agriculture Conservation Programs: The Role of Environmental Attitudes. *Journal of Socio-Economics,* 28: 335-349.

Marshall, G. 2004. From Words to Deeds: Enforcing Farmers' Conservation Cost-Sharing Commitments. *Journal of Rural Studies,* 20(2): 157-167.

Maruyama, M. 1985. The new logic of Japan's young generations. *Technological Forecasting and Social Change.*26(4): 351-364.

McCallum, C. 2003. *Identifying Barriers to Participation in Agri-Environmental Programs in Ontario.* Guelph: Christian Farmers Federation of Ontario.

Morito, B. 2002. *Thinking Ecologically – Environmental Thought, Values and Policy.* Halifax: Fernwood.

Morris, J., Mills, J. and Crawford, I. M. 2000. Promoting Farmer Uptake of Agri-Environmental Schemes: the Countryside Stewardship Arable Options Scheme. *Land Use Policy,* 17: 241-251.

Morris, C. and Potter, C.. 1995. Recruiting the New Conservationists: Farmers' Adoption of Agri-environmental Schemes in the U.K. *Journal of Rural Studies,* 11(1): 51-63.

Okey, B. 1998. *Application of Biodiversity Indicators in contrasting Farming Systems for Assessment of Agroecosystem Health.* PhD thesis, Department of Geography, University of Guelph.

Rogers, E. 1995. *Diffusion of Innovations.* N. Y.: Free Press.

Ryan, T. 1999. *The Clean Up Rural Beaches Program: Environmentalism in Action?* Guelph: University of Guelph MSc thesis.

Serman, N. 1999. *Socio-economic Factors which Influence the Farmer's Adoption Index Re-garding Soil and Water Conservation Practices in Southwestern Ontario.* Guelph: University of Guelph MSc thesis.

Smithers, J. and Furman, M. 2003. Environmental Farm Planning in Ontario: Exploring Participation and the Endurance of Change. *Land Use Policy,* 20(4): 343-356.

Statistics Canada. 2001. *Census of Agriculture.* http://www.statcan.ca/english/agcensus2001/index.htm. Retrieved October 6, 2004.

Statistics Canada. 2002. *Sharp Decline in Number of Farms in Ontario.* Retrieved June 4, 2002, from http://www.statcan.ca/english/agcensus2001/first/regions/farmon. htm#7.

Stern, P. C., Dietz, T., Kalof, L. and Guagnano, G. A. 1995. Values, beliefs and pro-environmental action: Attitude formation toward emergent attitude objects. *Journal of Applied Social Psychology* 25:1611–1636.

Stern, P. C. and Oskamp, S. 1987. Managing Scarce Environmental Resources. In Stokols, D. and Altman, I. (Eds), *Handbook of Environmental Psychology, Vol. 2.* New York: Wiley.

Stonehouse, D. P. 1996. A Targeted Policy Approach to Inducing Improved Rates of Conservation Compliance in Agriculture. *Canadian Journal of Agricultural Economics.* 44(2): 105-120.

Stonehouse, D. P., Hilts, S. G. and Blackburn, D. G.. May-June:1991. Soil Conservation Behavior and Attitudes Among Ontario Farmers. *Journal of Soil and Water Conservation,* 46(3): 215-219.

Tucker, M. and. Napier, T. L. 2002. Preferred Sources and Channels of Soil and Water Conservation Information Among Farmers in Three Midwestern US Watersheds. *Agriculture, Ecosystems and Environment,* 92: 297-313.

Van Donkersgoed, E. 2005. The Farmer's Stewardship Plan. *Corner Post #376.*

Wandel, J. 1995. An Analysis of Stability and Change in an Old Order Mennonite Farming System in Waterloo Region, Ontario. University of Guelph. Dept. of Geography, *MSc thesis.*

Wells, K. 2005. *Factors Affecting the Adoption of Best Management Practices in Canagagigue Creek.* Guelph: University of Guelph MSc thesis.

In: Watersheds
Editors: G. L. Kuhn and J. R. Emery
ISBN 978-1-62081-057-6

Chapter 6

ADAPTATION OF NATIONAL ELEVATION DATATSET TO ESTIMATE DEPRESSION STORAGE IN A GLACIATED WATERSHED*

Sarita Pachhai Karki[1,†] ***and Xixi Wang***[2,‡]
[1]Senior GIS Analyst I, Post, Buckley, Schuh and Jernigan, Inc (PBSandJ), Jacksonville, Florida, US
[2]Assistant Professor and Coordinator, Hydrology and Watershed Management Program, Department of Engineering and Physics, Tarleton State University, Stephenville, Texas, US

ABSTRACT

The construction of raised roads may modify the hydraulic connectivity of natural depression storages (e.g., wetlands) and create additional artificial storages. As a result, it would noticeably alter the natural hydrology of the inclusive watershed. Quantifying these storages is very important for watershed management that aims to increase water supply and mitigate flooding. This can be done by processing a digital elevation model (DEM) in a geographic information system (e.g., ArcGIS®) on the basis of the U.S. Geological Survey's 1.61-km by 1.61-

---

* A version of this chapter also appears in *Modeling Hydrologic Effects of Microtopographic Features*, edited by Xixi Wang, published by Nova Science Publishers, Inc. It was submitted for appropriate modifications in an effort to encourage wider dissemination of research.

† E-mail: spachhai@gnail.com.
‡ Tel. (254) 968-9164, E-mail: xxqqwang@gmail.com.

km quadrangle sections. However, DEMs with a spatial resolution finer than 30 m are not available for most watersheds. The objective of this study was to examine whether and how the 30-m National Elevation Dataset can be adapted to estimate potential depression storage capacity (PDSC), the total volume below the lowest point (i.e., breach) of the roads surrounding a section. The study was conducted in the Forest River watershed located in northeastern North Dakota. The results indicate that the estimation of PDSC tended to have larger errors for sections with a higher topographic relief than for sections with a lower relief. In addition, the estimation was sensitive to errors in the breach elevation values. Nevertheless, the PDSC for a section within the study watershed can be estimated as the multiplication of the value computed using the NED data by an adjustment coefficient of 1.05 to 1.3. One reasonable generalization of this study is that NED can be adapted to estimate PDSCs for watersheds where high-resolution DEMs are not available.

Keywords: DEM, GIS, depression, hydrology, PDSC, relief, roads

INTRODUCTION

The construction of roads and their associated drainage ditches can noticeably alter the natural hydrology of the inclusive watershed (Jones et al., 2000; Wigmosta and Perkins, 2001). This is particularly true for watersheds with gentle terrain, such as the Forest River watershed located in northeastern North Dakota (Figure 2.1). On the one hand, ditches may hydraulically connect natural depression areas (e.g., wetlands) with streams, increasing the contribution drainage area but shortening the distance required for a drop of water to travel through the overland into a stream (Magner et al., 2004). As a result, concentrated flow processes would become more prevalent and thus the time of concentration, the time required for a drop of water to travel from the most hydrologically remote point within a watershed to its outlet (Viessman and Lewis, 2003), would be reduced, leading to increased peak discharge (LaMarche and Lettenmaier, 2001). On the other hand, raised roads with a top above the adjacent ground levels could create artificial depression storages to store and attenuate the concentrated flows when conveyance structures (e.g., culverts and bridges) have insufficient capacities or are nonexistent (Duke et al., 2003). The artificial storages and the modified natural depression storages play a detention role, reducing the runoff contribution area, streamflow volume, and peak discharge. These two opposite effects of roads on watershed hydrology have been documented by a number of researchers (e.g., Harr et al.,

1975; Ziemer, 1981; King and Tennyson, 1984; Wright et al., 1990; Lewis et al., 2001).

Harr et al. (1975) conducted a paired watershed study in the Orgeon Coast Range and found that peak discharges for the combined fall and winter periods noticeably increased when roads occupied more than 12% of the watershed area. They also found that the flow hydrograph was shifted to have a delayed rising limb as a result of the detention function effects of artificial storages formed by the raised roads. In contrast, King and Tennyson (1984) showed a decrease in peak discharge through a paired watershed study conducted in north central Idaho. The authors stated that the noted decrease was the result of insufficient conveyance structures. Similarly, Ziemer et al. (1981) and Wright et al. (1990) conducted paired watershed studies in northwestern California and indicated no increase in peak discharge due to the construction of roads. The authors speculated that because the roads occupied only 5% of the watershed in size, which is much smaller than the 12% threshold value presented by Harr et al. (1975), the effects might not be detectable. Using the streamflow data collected at 14 gauging stations within the same watershed, Lewis et al. (2001) conducted a multiple regression analysis to study how the peak discharges and streamflow volumes depended on the areal extent of, and the number of stream crossings beneath, the roads. The results indicate an insignificant effect at a significance level of 0.05.

The inconsistent findings from the aforementioned studies as well as the others are unsurprising because effects of roads on the natural hydrology of a watershed depend on multiple factors, including: 1) the topology of road and stream networks (Wemple et al., 1996); 2) the road class (e.g., raised versus cut roads) and drainage type (e.g., relief culvert, mitre drain, and pushout drain; Croke and Mockler, 2001); 3) the road density, defined as the ratio of the total length of roads to the total drainage area of the watershed (Jones and Grant, 1996); 4) the topography (Veldhuisen and Russell, 1999); and 5) the weather (i.e.., wet versus dry) condition (Wemple and Jones, 2003). For a given weather condition, the first four factors affect how the natural storages would be modified and how many artificial storages would be created as a result of the construction of roads. When the density of raised roads is high, which is the case for the Forest River watershed, the artificial depression storages can greatly augment the natural depression storages, modifying the natural hydrology. Therefore, quantifying both the artificial and natural storages is very important for understanding consequences of such modifications as well as for making efforts to utilize the storages to increase water supply and mitigate flooding. FIPR (1998) explored the feasibility to use depressions from mining activities to recharge aquifers and increase water supply, whereas, Bolles and Wang (2003) studied how storages formed by raised roads could be utilized to temporarily store spring snowmelt runoff to reduce flooding.

The quantification can be implemented using a digital elevation model (DEM) and a geographic information system (e.g., ArcGIS®; Ormsby et al., 2001) on the basis of the U.S. Geological Survey (USGS) 1.61-km by 1.61-km quadrangle sections within a watershed. Herein, the key point is the availability and the required accuracy of DEM. The 30-m NED (National Elevation Dataset) is available for the conterminous United States and its territories. However, the question arised is whether the NED can be adapted to estimate the total (i.e., artificial plus natural depression) storages with sufficient accuracy. The answer to this question is seldom available in literature, though numerous studies (e.g., Garbrecht and Martz, 1997; Noman et al., 2003; Cochrane and Flanagan, 2005; Ziadat, 2007) have indicated that the spatial resolution of DEM affects the delineated watershed boundary and stream network as well as the estimated topographic parameters (e.g., slope, slope length, and relief). The objective of this study was to answer this question by comparing the storages estimated using the 30-m NED with those estimated using a 1-m LiDAR (Light Detection and Ranging) DEM for the Forest River watershed. To simplify the analysis, the total storage for a USGS quadrangle section, which is simply designated "section" for descriptive purposes, was defined as the total volume below the lowest top elevation of the roads surrounding the section.

Materials and Methods

The Forest River Watershed

The 2131 km^2 Forest River watershed, which is delineated as the USGS eight-digit hydrologic unit code of 09020308, covers portions of three North Dakota counties: Walsh, Nelson, and Grand Forks (Figure 2.1). The watershed comprises 823 USGS 1.61-km by 1.61-km sections, which form 22 townships that belong to the three counties. With the recently glaciated terrain, the watershed has a physiography of the Drift Prairie in the upper (i.e., western) portion and the Red River Valley Lake Plain in the lower (i.e., eastern) portion (Stoner et al., 1993; LeFever et al., 1999).

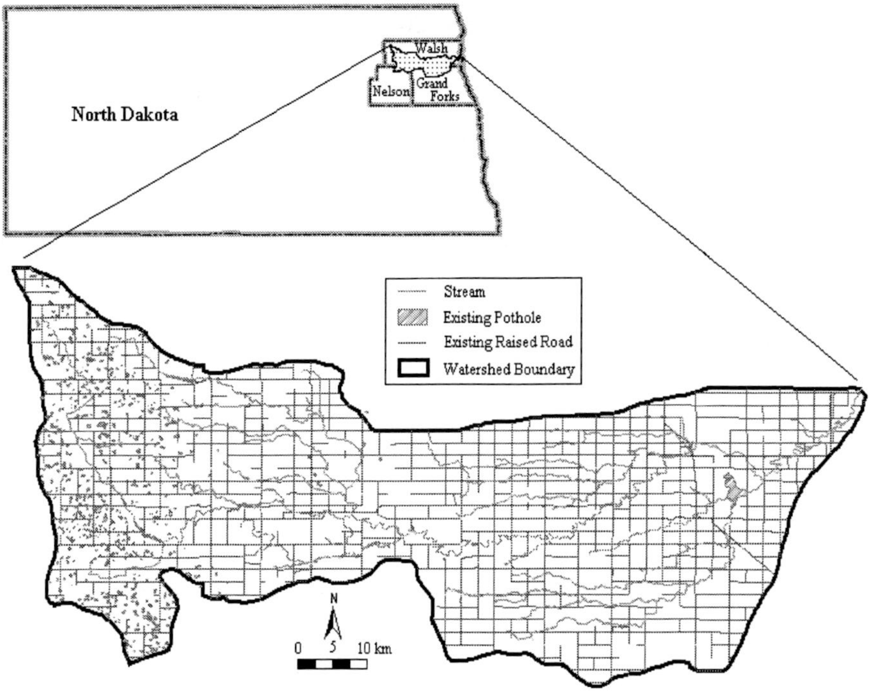

Figure 2.1. Map showing the location and boundary of the Forest River watershed, superimposed by the stream network and existing potholes presented by the U.S. Geological Survey's National Hydrography Dataset (NHD).

The Drift Prairie physiographic area is characterized by rolling hills and numerous prairie potholes (Mark, 1988; Dredge, 2000). Prairie potholes are small ponds and wetlands surrounded by marshy borders that occupy closed basins and do not contribute runoff to streams. Based on the National Hydrography Dataset (NHD; USGS, 2001), the prairie potholes account for approximately 0.9% (19.4 km^2) of the Forest River watershed in size. NHD is a comprehensive set of digital spatial data that contain the information about surface water features such as lakes, ponds, streams, rivers, springs, and wells. Along with the potholes that have been drained, dredged, filled, and leveled since colonization (USGS, 2001), which are not included in NHD, the watershed used to have about 80 km^2 prairie potholes as estimated by Gleason and Euliss (1998). In contrast, the Red River Valley Lake Plain physiographic area has a very low topographic relief and thus few ponds and lakes, as indicated by the information presented by the NHD. The potholes are natural depression storages for runoff. On the other hand, the existing raised roads, with a total length of approximately 2070 km and constructed along the section and quarter-section lines in the Forest River watershed (Figure 2.1; USGS, 1996), created a large amount of artificial storages (Duke et al., 2003). Because of the combined effects of the natural and artificial storages, the runoff contribution area accounts for only 84% (i.e., 1785 out of 2131 km^2) of the watershed in size.

Roads have been cited as one of the major factors inducing the altered hydrology in watersheds (e.g., USEPA, 2006). A raised road has a top elevation higher than that of its adjacent ditches and/or fields, modifying the direction, velocity, and conveyance process of overland flow (Wemple et al., 1996; Jones et al., 2000; Tague and Band, 2001). The installation of culverts at locations where the overland flow tends to accumulate can facilitate the conveyance process, whereas, the absence of culverts would cause the flow to be ponded at these locations. The ponded water is ultimately lost to evaporation and infiltration. Hereafter, these locations are designated "breaches" for descriptive purposes. For a given section, the volume below the breach elevation is designated the "potential depression storage capacity (PDSC)," the total volume of water that can be stored by the natural and artificial storages assuming that no culvert is installed. That is, for the section, its PDSC is greater than its actual storage capacity when one culvert or more exists. Also, the PDSC excludes the volumes below the annual average water surface elevations of the potholes. These volumes are perennially occupied by water and not available for storing additional overland flow.

Datasets Used

Three national datasets, namely NHD, DLG (Digital Line Graph), and NED, were downloaded from the USGS website http://edcftp.cr.usgs.gov/pub/data. As mentioned above, NHD is a comprehensive set of digital data that contain the information about surface water features such as lakes, ponds, streams, rivers, springs, and wells (USGS, 2001). This study used the NHD data layer to define the boundary of, and to identify the stream network and potholes in, the Forest River watershed.

DLG, with a spatial scale of 1:24,000, is the digital representation of cartographic information (USGS, 1996). DLG consists of nine data layers, namely: 1) Public Land Survey System (PLSS) that provides the information on township, range, and section line; 2) boundaries of state, county, city, and other national and state lands (e.g., forests and parks); 3) transportation (e.g., roads, trails, and railroads) and utility supply network (e.g., pipelines and transmission lines); 4) hydrology, such as flowing water, standing water, and wetlands; 5) hypsography, such as contours and supplementary spot elevations; 6) non-vegetative features of lava, sand, and gravel; 7) survey control and bench markers defined by horizontal and vertical positions; 8) manmade features (e.g., building); and 9) vegetative surface cover, such as woods, scrub, orchards, vineyards, and vegetative features associated with wetlands. In this study, the PLSS data layer was used to define the section lines within the Forest River watershed, whereas, the transportation layer was used to identify types of the roads existed along the section lines. For a given section, the elevations at the four corners and four mid points of the section lines were extracted from the hypsography data layer, giving eight known road surface elevations. Because the elevations at other locations along the roads are not available, this study assumed that the road surfaces have linear slopes between any two of the adjacent known-elevation points. Also, this study assumed that the minimal value of these eight known elevations be the breach elevation of the section.

In order to verify these assumptions, three sections were randomly selected for the Forest River watershed. A differential global positioning system (DGPS), which was calibrated to have a horizontal accuracy of less than 1 m and a vertical accuracy of less than 10 cm (Wikimedia Foundation Inc., 2006), was used to survey more than one million points along the four roads surrounding each of the selected sections (Pachhai, 2005; K. William, Research Scientist in the Energy and Environmental Research Center at University of North Dakota, personal communication, 2006). The profiles (not shown), plotted using these surveyed elevations, indicate that for each selected section, the lowest road surface location coincides with the breach that was assumed to be the lowest one of the eight known-elevation points.

NED was designed to provide National elevation data in a seamless form with a consistent datum, elevation unit, and projection (USGS, 2006). NED has a resolution of one arc-second (approximately 30 m) for the conterminous United States, Hawaii, Puerto Rico and the island territories and a resolution of two arc-second for Alaska. The data sources used to develop NED have a variety of elevation units, horizontal units, and map projections. As a result, NED has an accuracy that varies from one area to another. The NED data for the Forest River watershed has a vertical accuracy of between 1 and 2 m (Holmes et al., 2000; Gesch et al., 2002).

In the early spring of 2004, when the vegetations (in particular, trees and tall grasses) did not germinate and thus had a minimal coverage of the ground,

Sanborn (http://www.sanborn.com) collected the evaluation data for the Forest River watershed using a LiDAR system that utilizes an airborne laser beam as the sensing carrier (Wehr and Lohr, 1999) and can achieve a very high accuracy when the laser pulses are least prevented from reaching the ground surface (Lillesand et al., 2004). Subsequently, Sanborn conducted a post processing to remove buildings, automobiles, and vegetation stubs from the collected data. The processed data were further analyzed and resampled to develop a 1-m elevation dataset, which was designated the "LiDAR DEM" for descriptive purposes. To assess the vertical accuracy of the LiDAR DEM, the ground elevations for 106 locations that were randomly selected across the watershed were surveyed using the aforementioned DGPS (K. William, Research Scientist in the Energy and Environmental Research Center at University of North Dakota, personal communication, 2006) and compared with the corresponding values presented by the LiDAR DEM. The assessment indicated that the LiDAR DEM has a vertical accuracy of between 0.03 and 0.46 m (Pachhai, 2005). For a given location within the study watershed, the absolute difference between the elevation presented by the LiDAR DEM and that presented by the NED data can be very large (Figure 2.2). The difference tends to be smaller for the Red River Valley Lake Plain than for the Drift Prairie, with the largest differences for the areas adjacent to the streams.

Study Methods

The values for PDSC estimated using the LiDAR DEM were presumed to be accurate and thus taken as the references for assessing the values for PDSC estimated using the NED data. Pachhai (2005) indicated that for a section, the difference of the estimated PDSC values using these two datasets depends on the topographic relief of the section. For the purpose of fair comparison, the 823 sections within the Forest River watershed were classified into three categories. The Category–I sections are characterized by having a low topographic relief, whereas, the Category–III sections are characterized by having a high topographic relief. The Category–II sections have a moderate topographic relief. The index that is commonly used to measure topographic relief is defined as the difference of the highest elevation than the lowest one within the area of interest (herein, section). Obviously, this index is a function of the spatial resolution (i.e., cell size) of DEM (Gao, 1997; Kienzle, 2004): a DEM with a higher resolution tends to give a greater value of the index than the one with a lower resolution, and vice versa. For this reason, the index was not used to classify the sections in this study because the LiDAR DEM and NED have distinctly different spatial resolutions, probably giving incompatible classifications.

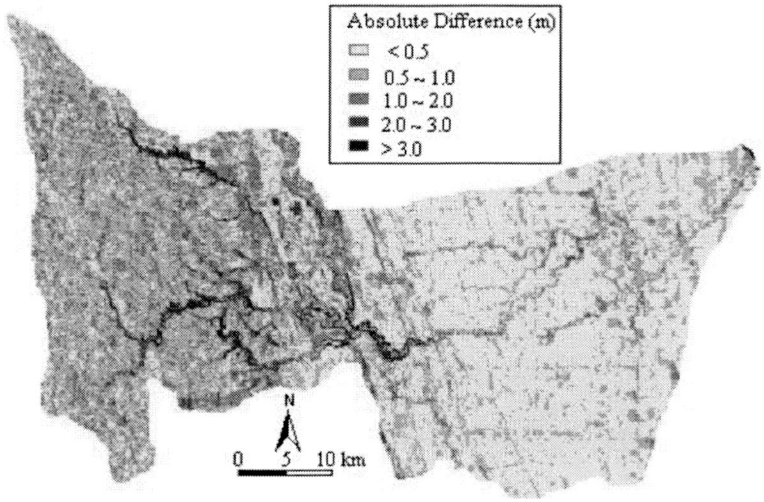

Figure 2.2. Map showing the absolute differences between the elevations presented by the LiDAR DEM and those presented by the U.S. Geological Survey's National Elevation Dataset (NED).

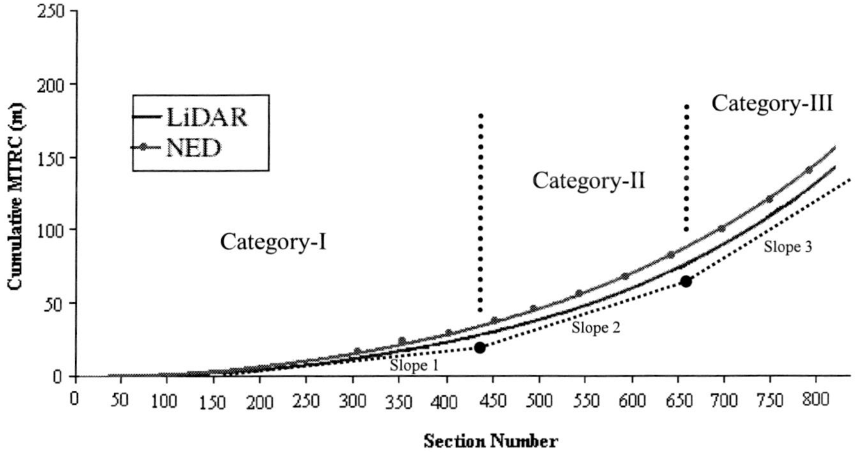

Figure 2.3. Plot showing the cumulative modified topographic relief classifier (MTRC) computed using the 1-m LiDAR DEM and 30-m National Elevation Dataset (NED). Based on the MTRC, the sections in the Forest River watershed were classified into Category-I, -II, and –III.

Instead, this study developed and used a new index, which is defined as the difference of the average elevation of a section minus the median elevation of the same section. To be differentiated from the aforementioned conventional index, this newly developed index was designated MTRC (modified topographic relief classifier) for descriptive purposes. The MTRC is almost independent of the resolution of DEM, as indicated by that the values of MTRC computed using the LiDAR DEM and those computed using the NED data for the sections within the study watershed are visually same (Figure 2.3). This was further verified by a paired t-test, which indicated that the two estimated values for a given section are statistically identical at a significance level of 0.05 (p-value = 0.786). Thus, this study used the MTRC to classify the 823 sections by taking the points (Figure 2.3), at which the slope of the curves has a noticeable change, as the breakpoints for the three categories. A Category–I section was defined to have a MTRC value less than 0.35 m, whereas, a Category–III section was defined to have a MTRC value greater than 0.80 m. A Category–II section was defined to have a MTRC value between 0.35 and 0.80 m.

Along each of the 17 south-north transects that are about 3 km apart, three to six sections were randomly selected to compute the values of PDSC using the LiDAR DEM and the NED data, respectively. The computation was implemented section by section using the ArcGIS® "Surface Volume" function, with the "Plane Height Below" specified as the breach elevation of a selected section. The detailed description of this function can be found in Ormsby et al. (2001). Subsequently, visualization plots and maps were generated to compare the values for PDSC computed using the LiDAR DEM data with those computed using the NED data.

The breach elevation value presented by the hypsography data layer of DLG has an accuracy of ±0.45 m (M. Bearden, Research Geographer, the U.S. Geological Survey, personal communication, 2005). To assess the sensitivity of PDSC to errors included in a breach elevation value, one Category–I section, one Category–II section and one Category–III section were randomly selected. For each of the three sections, the values for PDSC with the "Plane Height Below" specified as ±0.15 and ±0.45 of the breach elevation value were computed using both the LiDAR DEM and NED data. The computed values for PDSC for each section were plotted separately to examine the sensitivity.

RESULTS AND DISCUSSION

Classification of the Sections

Base on the MTRC, 85% (i.e., 703 out of 823) of the sections had a consistent classification using the LiDAR DEM and the NED data. However, each of the remaining 120 sections was assigned into different categories (Table 2.1 and Figure 2.4). This indicated that the MTRC is a powerful classifier irrespective of the spatial resolution of DEM. A close examination revealed that most of the misclassified sections are adjacent to the streams (Figure 2.4 versus Figure 2.1), portions of which are located in the floodplains and could be inundated during flooding season (i.e., spring) but may become dry for the other seasons. The elevations presented both by the LiDAR DEM and the NED data are likely to be the water surfaces at the time when the data were collected (Lillesand et al., 2004; USGS, 2006). Thus, the misclassification of these sections might be because the water surfaces at the time when the LiDAR data were collected were very different from those at the time when the data used to develop NED were collected. Another possible reason is that the dense tall vegetations (e.g., trees) along, and in the vicinity of, the streams caused large errors in the collected data (Hopkinson et al., 2004), as indicated by the larger differences between the elevations presented by the LiDAR DEM data and those presented by the NED data (Figure 2.2). For these reasons, the misclassified sections were excluded from further analysis.

Computed PDSC

In this study, 44 Category–I, 30 Category–II, and 24 Category–III sections were randomly selected to compute the values for PDSC using the LiDAR DEM and NED data. Averaged across the 98 sections, the value for PDSC was estimated to be 2.7×10^5 m^3 per section using the NED data and 3.3×10^5 m^3 per section using the LiDAR DEM data. For all three categories, the values for PDSC computed using the NED data were smaller than the corresponding values computed using the LiDAR DEM data, with a greatest difference of 37.5% for the Category–III sections and a smallest difference of 7.7% for the Category–II sections (Figure 2.5). The difference for the Category–I sections was 19.4%.

With a higher spatial resolution, the LiDAR DEM presents the information that can be used to accurately characterize the minor terrain features such as draws and spurs. A draw is a less developed stream course, whereas, a spur is a short, continuous sloping line of higher ground, and is often formed by two roughly parallel streams (DOD, 1999). However, these minor terrain features are smoothed out in the 30-m NED data. As a result, the storages created by these minor features could not be quantified using the NED data, leading to the underestimation of PDSC. The Category–III sections have more minor features than the Category–I and Category–II sections (Stoner et al., 1993). Thus, the difference of the values for PDSC computed using these two DEM datasets was largest for the Category–III sections (Figure 2.6). In addition, because the Category–I sections have a low topographic relief, a small discrepancy of the elevations provided by the LiDAR DEM and the NED data (Figure 2.2) can result in a large difference of the computed storage areas (Figure 2.7) and thus in a large difference of the computed values for PDSC. In contrast, the Category–II sections have fewer minor terrain features and the PDSC is least sensitive to the small discrepancy of the elevations presented by these two DEM datasets (Figure 2.8).

Table 2.1. The classes of the 823 sections in the Forest River watershed

Class	Definition [1]	Number of Sections	Consistent for the LiDAR DEM and the NED data? [2]
Category–I	MTRC < 0.35 m	504	Yes
Category–II	0.35 m ≤ MTRC ≤ 0.80 m	99	Yes
Category–III	MTRC > 0.80 m	100	Yes
Misclassified		120	No

[1] MTRC is the modified topographic relief classifier, defined as the average elevation of a section minus the median elevation of the section.

[2] The LiDAR DEM is the 1-m Light Detection and Ranging Digital Elevation Model, whereas, the NED data are the U.S. Geological Survey's 30-m National Elevation Dataset.

However, for the Category–III sections, the values for PDSC computed using the LiDAR DEM were well correlated with the corresponding values computed using the NED data, as indicated by a high coefficient of determination (R^2) of 0.95 (Figure 2.9c). In contrast, the computed values for PDSC using these two DEM datasets exhibited a moderate inconsistency for the Category–I and Category–II sections (Figures 9a and 9b), indicating a relatively greater sensitivity to data resolution. This is consistent with the findings of Noman et al. (2003), Hill and Neary (2005), and Martin et al. (2008). The results showed that for watersheds where a high-resolution LiDAR DEM is not available, NED can be used to estimate the PDSCs with an adjustment coefficient of 1.05 to 1.3. That is, the PDSC for a section can be estimated as the multiplication of the value computed using the NED data by the adjustment coefficient.

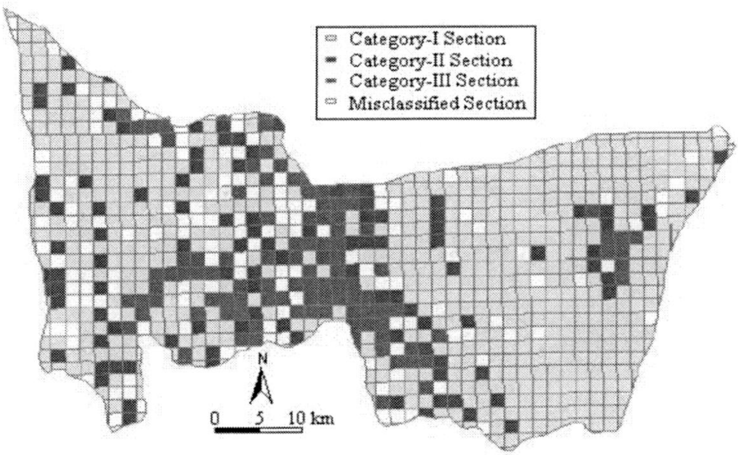

Figure 2.4. Map showing the classification consistency using the 1-m LiDAR DEM and the U.S. Geological Survey's 30-m National Elevation Dataset (NED).

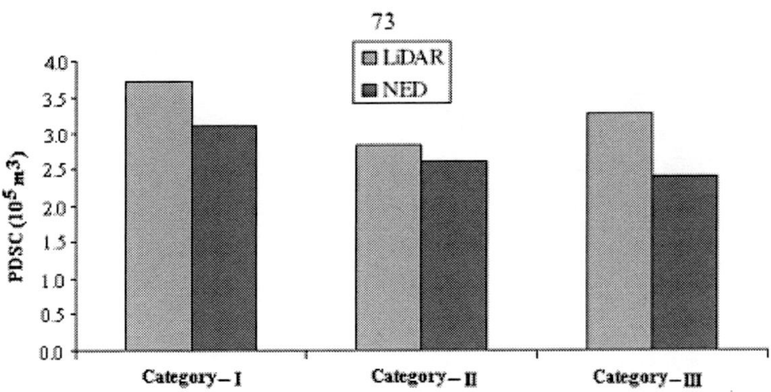

Figure 2.5. The section-averaged values for potential depression storage capacity (PDSC) computed using the 1-m LiDAR DEM versus those computed using the 30-m National Elevation Dataset (NED) for selected Category-I, -II, and –III sections.

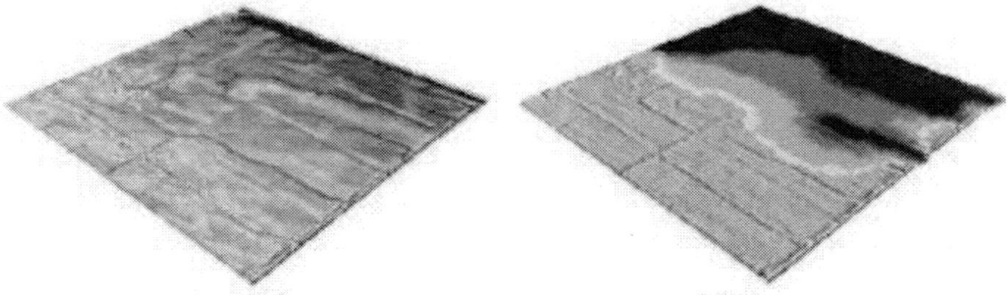

(a) (b)

Figure 2.6. The potential depression storage area for a typical Category–III section determined using the (a) 1-m LiDAR DEM and (b) 30-m National Elevation Dataset (NED). The areas in blue color are water.

(a) (b)

Figure 2.7. The potential depression storage area for a typical Category–I section determined using the (a) 1-m LiDAR DEM and (b) 30-m National Elevation Dataset (NED). The areas in blue color are water.

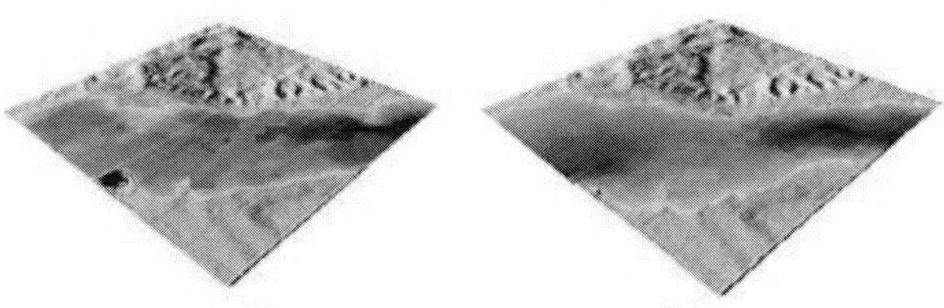

(a) (b)

Figure 2.8. The potential depression storage area for a typical Category–II section determined using the (a) 1-m LiDAR DEM and (b) 30-m National Elevation Dataset (NED). The areas in blue color are water.

A coefficient approaching to the upper bound should be applied to Category–III sections, whereas, a coefficient close to the lower bound is more appropriate for Category–II sections. A coefficient around the midpoint of the range (i.e., 1.15) is probably a best choice for Category–I sections.

Sensitivity of PDSC Estimation to Breach Elevation Value

For a given section, errors in the breach elevation value affect the estimation of PSDC (Figure 2.10). As expected, a positive error would result in the overestimation of PSDC, whereas, a negative error would result in the underestimation of PSDC. In addition, the estimation of PSDC is more sensitive to negative than positive errors. Nevertheless, the sensitivity is independent of data resolution, as indicated by that for a given variation of the breach elevation value (e.g., +0.15 m), the corresponding variation of the values for PSDC computed using the LiDAR DEM is almost identical to that of the values for PSDC computed using the NED data (Figure 2.10). This is probably because the water at 0.45 m below the breach elevation may be deep enough to fill the draws, minimizing influences of the minor terrain features on the estimation of PSDC. Compared with the Category–II and –III sections, the Category–I section was determined to be most sensitive to errors in the breach elevation value. Again, because the Category–I section has a low topographic relief, a small discrepancy of the elevation value can result in a large difference of the estimated storage area.

Conclusion

This study developed MTRC, which is computed as the difference of the average elevation of a section from the median elevation of this same section, and showed that MTRC is independent of the spatial resolution of DEM. Using this newly developed index, this study classified the 823 sections of the Forest River watershed, which is located in northeastern North Dakota, into three categories. The Category–I sections have a MTRC value less than 0.35 m, whereas, the Category–III sections have a MTRC value greater than 0.80 m. The Category–II sections have a MTRC value between 0.35 and 0.80. A section with a larger MTRC value tends to have a greater topographic relief than the one with a smaller MTRC value. Subsequently, this study evaluated the feasibility to estimate PDSCs using the available NED data. The evaluation was implemented by comparing the values for PDSC computed using a 1-m LiDAR DEM with those computed using the 30-m NED data for 98 sections that were randomly selected from these three categories.

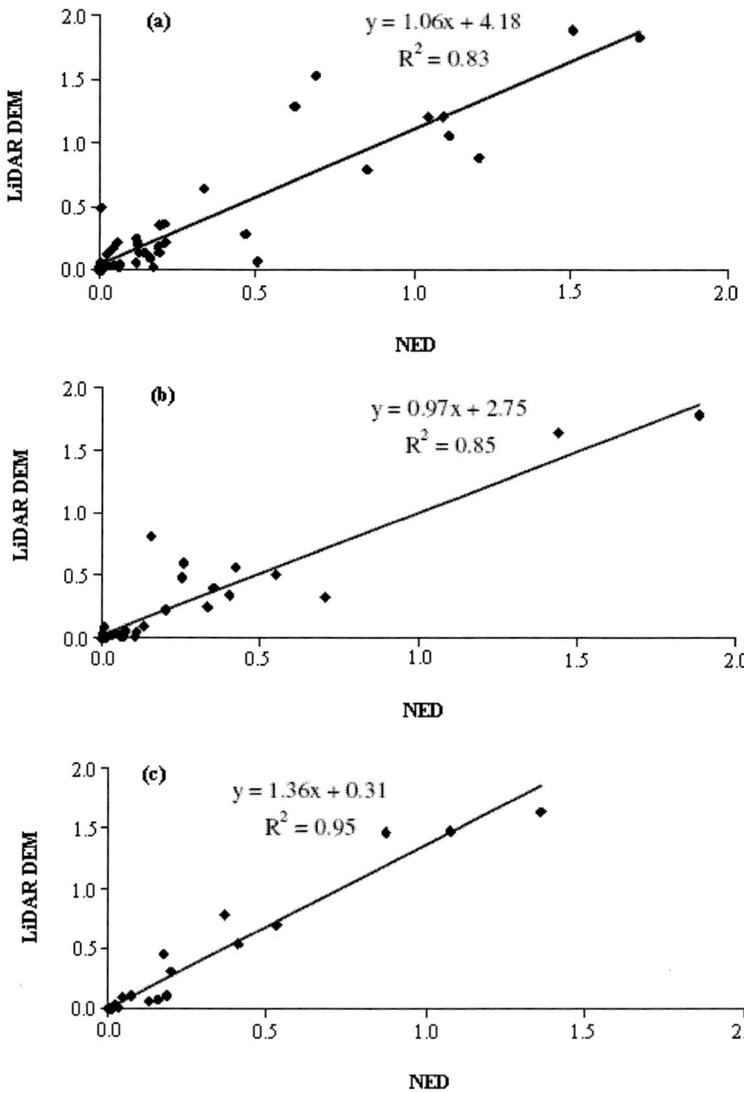

Figure 2.9. Plot showing the values for potential depression storage capacity (PDSC), in 10^6 m^3, computed using the 1-m LiDAR DEM versus those computed using the 30-m National Elevation Dataset (NED) for randomly selected (a) Category-I, (b) Category-II, and (c) category-III, sections.

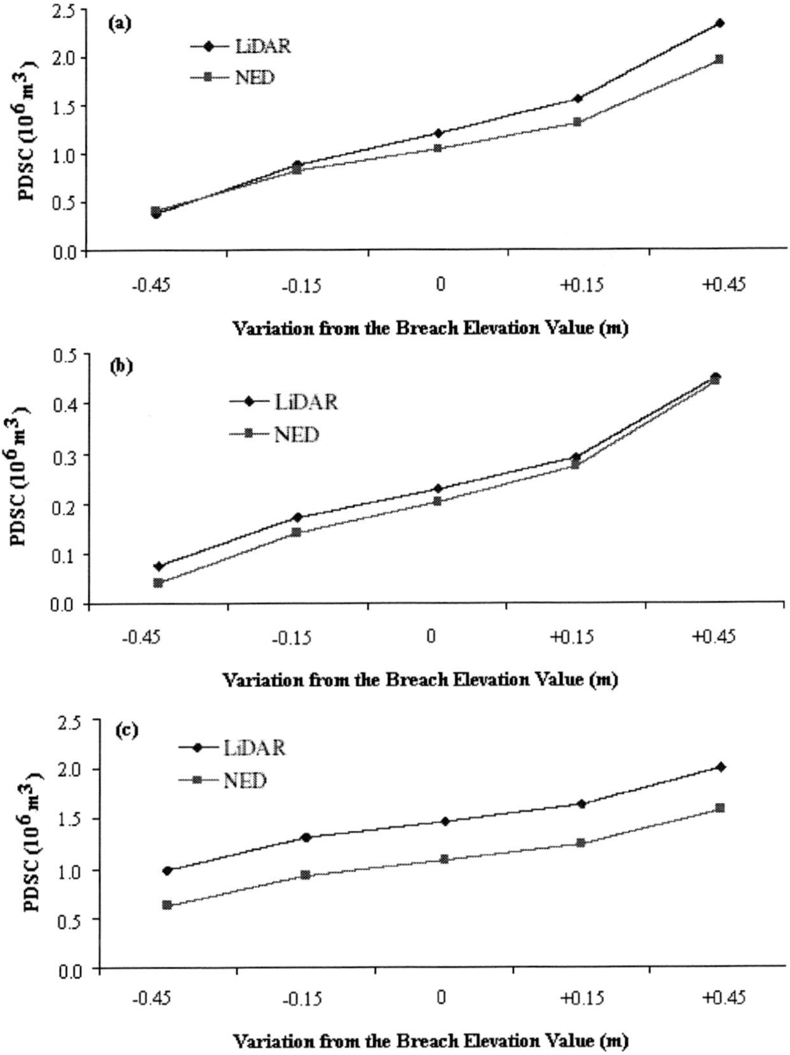

Figure 2.10. Sensitivity of the estimation of potential depression storage capacity (PDSC) to breach elevation value as evaluated for randomly selected (a) Category–I, (b) Category–II, and (c) Category–III, sections.

For a section, compared with the NED data, the LiDAR DEM tended to give a greater value for PDSC. Across the study watershed, the average value for PDSC was estimated to be 2.7×10^5 m^3 per section using the NED data and 3.3×10^5 m^3 per section using the LiDAR DEM. The difference between the values for PDSC computed using the LiDAR DEM and those computed using the NED data was largest for the Category–III sections, followed by the Category–I and then Category–II sections. The estimation of PDSC is more sensitive to errors in the breach elevation value for a Category–I section than for Category-II and –III sections. Nevertheless, for all three category sections, the values of PDSC computed using the LiDAR DEM were well correlated with the values computed using the NED data ($R^2 \geq 0.83$). The results indicate that the PDSCs for the study watershed can be estimated as the multiplications of the values for PDSC computed using the NED data by an adjustment coefficient of 1.05 to 1.3. A coefficient approaching to the upper bound should be applied to the Category–III sections, whereas, a coefficient close to the lower bound is appropriate for the Category–II sections. A coefficient around the midpoint of the range (i.e., 1.15) is probably a best choice for the Category–I sections. One reasonable generalization is that NED can be adapted to estimate PDSCs for watersheds where a high-resolution LiDAR DEM is not available.

References

Bolles, B.A., and Wang, X. (2003). Distributed basinwide storage for flood mitigation. In *Proceedings of the 20th Annual Red River Basin Land and Water International Summit Conference*. Winnipeg, Manitoba, Canada.

Cochrane, T.A., and Flanagan, D.C. (2005). Effect of DEM resolutions in the runoff and soil loss predictions of the WEPP watershed model. *Transactions of the ASAE* 48(1): 109 – 120.

Croke, J., and Mockler, S. (2001). Gully initiation and road-to-stream linkage in a forested catchment, southeastern Australia. *Earth Surface Processes and Landforms* 26: 205 – 217.

DOD (Department of Defense). (1999). Map Reading and Land Navigation. Washington D.C., Apple Pie Publishers.

Dredge, L. A. (2000). Age and origin of upland block fields on Melville Peninsula, eastern Canadian Arctic. *Geografiska Annaler* 82A(4): 443 – 454.

Duke, G. D., Kienzle, S.W., Johnson, D. L., and Byrne, J. M. (2003). Improving overland flow routing by incorporating ancillary road data into digital elevation models. *Journal of Spatial Hydrology* 3(2): 1 – 27.

FIRP (Florida Institute of Phosphate Research). (1998). Evaluation of the feasibility of water storage reservoir on mined lands to meet future agricultural, potable, industrial, and public water supply demands. WWW document, http://www.fipr.state.fl.us/ar97/ar97wate.htm.

Gao, J. (1997). Resolution and accuracy of terrain representation by grid DEMs at a micro-scale. *International Journal of Geographical Information Science* 11(2): 199 – 212.

Garbrecht, J., and Martz, L. W. (1997). TOPAZ: *An Automated Digital Landscape Analysis Tool for Topographic Evaluation, Drainage Identification, Watershed Segmentation, and Subcatchment Parameterization*: Overview. Durant, Oklahoma, the U.S. Department of Agriculture Agricultural Research Service ARS-NAWQL 95 – 1.

Gesch, D., Oimoen, M., Greenlee, S., Nelson, S., Steuck, C., and Tyler, D. (2002). The national elevation dataset. *Photogrammetric Engineering and Remote Sensing* 68(1): 5 – 15.

Gleason, R.A., and Euliss, N.H. Jr. (1998). Sedimentation of prairie wetlands. *Great Plains Research* 8(1): 97 – 112.

Harr, D. H., Harper, W.C., and Krieger, J.T. (1975). Changes in storm hydrographs after road building and clear-cutting in the Oregon Coast Range. *Water Resources Research* 11(3): 436 – 444.

Hill, A. J., and Neary, V.S. (2005). Factors affecting estimates of average watershed slope. *Journal of Hydrologic Engineering* 10(2): 133 – 140.

Holmes, K.W., Chadwick, O. A., and Kyriakidis, P. C. (2000). Error in a USGS 30-meter digital elevation model and its impact on terrain modeling. *Journal of Hydrology* 233: 154 – 173.

Hopkinson, C., Chasmer, L.E., Zsigovics, G., Greed, I.F., Sitar, M., Treitz, P., and Maher, R. V. (2004). Errors in LiDAR ground elevation and wetland vegetation height estimates. *International Archives of Photogrammetry, Remote Sensing and Spatial Information Services* XXXVI–8/W2.

Jones, J. A., and Grant, G. E. (1996). Peak flow responses to clear-cutting and roads in small and large basins, Western Cascades, Oregon. *Water Resources Research* 32(4): 959 – 974.

Jones, J. A., Swanson, F. J., Wemple, B. C., and Snyder, K. U. (2000). Effects of roads on hydrology, geomorphology, and disturbance patches in stream networks. *Conservation Biology* 14(1): 76 – 85.

Kienzle, S. (2004). The effect of DEM raster resolution on first order, second order and compound terrain derivatives. *Transactions in GIS* 8(1): 83 – 111.

King, J. G., and Tennyson, L. C. (1984). Alteration of streamflow characteristics following road construction in north central Idaho. *Water Resources Research* 20(8): 1159 – 1163.

LaMarche, J. L., and Lettenmaier, D. P. (2001). Effects of forest roads on flood flows in the Deschutes River, Washington. *Earth Surface Processes and Landforms* 26: 115 – 134.

LeFever, J. A., Bluemle, J. P., and Waldkirch, R. P. (1999). Flooding in the Grand Forks–East Grand Forks, North Dakota and Minnesota Area. Bismarck, North Dakota, North Dakota Geological Survey Educational series No 25.

Lewis, J., Mori, S.R., Keppeler, E.T., and Ziemer, R.R. (2001). Impacts of logging on storm peak flows, flow volumes and suspended sediment loads in Caspar Creek, California. In *Proceedings of Land Use and Watersheds: Human Influence on Hydrology and Geomorphology in Urban and Forest Areas* (ed. M S Wigmosta and S J Burges). Washington D.C., American Geophysical Union: 127 – 144.

Lillesand, T.M., Kiefer, R.W., and Chipman, J.W. (2004). Remote Sensing and Image Interpretation (5th ed.). New York, John Wiley and Sons Ltd.

Magner, J.A., Payne, G.A., and Steffen, L.J. (2004). Drainage effects on stream nitrate-N and hydrology in south-central Minnesota (USA). *Environmental Monitoring and Assessment* 91(1 – 3): 183 – 198.

Mark, D.M. (1988). Network models in geomorphology. In Proceedings of Modeling Geomorphological Systems (ed. Anderson M G), New York, John Wiley and Sons Ltd. 73 – 97.

Martin, Y.E., Valeo, C., Tait, M., and Johnson, E. (2008). Impacts of centimetre-scale landscape roughness on depression storage, runoff and sediment transport. In *Proceedings of American Geophysical Union Fall Meeting 2008*. Washington D.C., American Geophysical Union, Abstract #H51D-0848.

Noman, N.S., Nelson, E.J., and Zundel, A.K. (2003). Improved process for floodplain delineation from digital terrain models. *Journal of Water Resources Planning and Management* 129(5): 427 – 436.

Ormsby, T., Napoleon, E., Burke, R., Groessl, C., and Feaster, L. (2001). *Getting to Know ArcGIS Desktop*. Redlands, Calofornia, ESRI Press.

Pachhai, S. (2005). Comparison Between 1-m LiDAR and 30-m NED Digital Elevation Models for Temporary Floodwater Storage Estimations in the Red River of the North Basin. Grand Forks, North Dakota, University of North Dakota, Geography Department, Master Thesis.

Stoner, J.D., Lorenz, D.L., Wiche, G.J., and Goldstein, R.M. (1993). Red River of the North Basin, Minnesota, North Dakota, and South Dakota. *Water Resources Bulletin* 29(4): 575 – 615.

Tague, C., and Band, L. (2001). Simulating the impact of road construction and forest harvesting on hydrologic response. *Earth Surface Processes and Landforms* 26(2): 135 – 152.

USEPA (U.S. Environmental Protection Agency). (2006). Hydrology. WWW document, http://www.epa.gov/owow/watershed/wacademy/wam/hydrology.html.

USGS (U.S. Geological Survey). (1996). US GeoData: Digital line graphs. WWW document, http://erg.usgs.gov/isb/pubs/factsheets/fs07896t.pdf.

USGS (U.S. Geological Survey). (2001). National Hydrography Dataset. WWW document, http://nhd.usgs.gov.

USGS (U.S. Geological Survey). (2006). National Elevation Dataset. WWW document, http://ned.usgs.gov/Ned/about.asp.

Veldhuisen, C., and Russell, P. (1999). Forest Road Drainage and Erosion Initiation in Four West-Cascade Watersheds. TFW Effectiveness Monitoring Report TFW-MAG1-99-001.

Viessman, W. Jr., and Lewis, G.L. (2003). *Introduction to Hydrology* (5th ed.). Prentice Hall, Upper Saddle River, NJ.

Wehr, A., and Lohr, U. (1999). Airborne laser scanning: An introduction and overview. *ISPRS Journal of Photogrammetry and Remote Sensing* 54(2): 68 – 82.

Wemple, B.C., and Jones, J.A. (2003). Runoff production on forest roads in a steep, mountain catchment. *Water Resources Research* 39(8): 1220 – 1236.

Wemple, B.C., Jones, J.A., and Grant, G.E. (1996). Channel network extension by logging roads in two basins, Western Cascades, Oregon. *Water resources Bulletin* 32(6): 1195 – 1207.

Wigmosta, M. S., and Perkins, W.A. (2001). Simulating the effects of forest roads on watershed hydrology. In *Proceedings of Land Use and Watersheds: Human Influence on Hydrology and Geomorphology in Urban and Forest Areas* (ed. M S Wigmosta and S J Burges). Washington D.C., American Geophysical Union: 127 – 144.

Wikimedia Foundation Inc. (2006). Differential GPS. WWW document, http://en.wikipedia.org/wiki/Differential_GPS.

Wright, K.A., Sendek, K.H., Rice, R.M., and Thomas, R.B. (1990). Logging effects on streamflow: Storm runoff at Caspar Creek in northwestern California. *Water Resources Research* 26(7): 1657 – 1667.

Ziadat, F.M. (2007). Effect of contour intervals and grid cell size on the accuracy of DEMs and slope derivatives. *Transactions in GIS* 11(1): 67 – 81.

Ziemer, R.R. (1981). Storm flow response to road building and partial cutting in small streams of northern California. *Water Resources Research* 17(4): 907 – 917.

INDEX

A

B

E

F

P

Q

R

T

U

V

W

Y

Z